AGAINST THE GRAIN
Agri-environmental Reform in the United States and the European Union

To my parents

AGAINST THE GRAIN
Agri-environmental Reform in the United States and the European Union

Clive Potter

Environment Department, Wye College, University of London, Ashford, UK

CAB INTERNATIONAL

CAB INTERNATIONAL
Wallingford
Oxon OX10 8DE
UK

Tel: +44 (0)1491 832111
Fax: +44 (0)1491 833508
E-mail: cabi@cabi.org

CAB INTERNATIONAL
198 Madison Avenue
New York, NY 10016-4314
USA

Tel: +1 212 726 6490
Fax: +1 212 686 7993
E-mail: cabi-nao@cabi.org

A catalogue record for this book is available from the British Library,
London, UK.

Library of Congress Cataloging-in-Publication Data
Potter, Clive, 1959–
      Against the grain : agri-environmental reform in the United States
   and the European Union / Clive Potter.
         p.   cm.
      Includes bibliographical references (p.   ) and index.
      ISBN 0-85199-228-5 (alk. paper)
      1. Agriculture– –Environmental aspects– –Government policy– –United
   States.   2. Agriculture– –Environmental aspects– –Government policy– –
   European Union countries.   3. Agriculture and state – – United States.
   4. Agriculture and state– –European Union countries.
   5. Environmental policy– –United States.   6. Environmental policy– –
   European Union countries.   I. Title.
   S589.755.P68   1998
   363.7– –dc21   97–33664
                                                            CI

ISBN 0 85199 228 5

Typeset in Optima and Photina MT by Advance Typesetting Ltd, Oxford
Printed and bound by Biddles Ltd, King's Lynn and Guildford

Contents

Abbreviations   vi

Foreword   ix
David E. Ervin

Introduction   1

1   Engines of Destruction?   9

2   The Pressures for Reform   35

3   The Conservation New Deal   61

4   Agricultural Stewardship in the UK   82

5   The Defence of Green Europe   105

6   Agricultural Liberalization and the Double Dividend   128

7   The Environmental Reform of Farm Policy?   154

References   163

Index   187

Abbreviations

AAA	Agricultural Adjustment Administration
AAPS	Arable Area Payments Scheme
ACP	Agricultural Conservation Program
AEP	Agri-Environmental Programme
AFB	American Farm Bureau
AFT	American Farmland Trust
AMS	Aggregate Measure of Support
ARP	Acreage Reduction Program
ASCS	Agricultural Stabilization and Conservation Service
BGMS	Broads Grazing Marsh Scheme
CAP	Common Agricultural Policy
CARPE	Common Agricultural and Rural Policy for Europe
CC	Countryside Commission
CCW	Countryside Council for Wales
CEAFG	Council of Environmental Advisors to the Federal Government
CEEC	Central and Eastern European country
CLA	Country Landowners' Association
COPA	Committee of Professional Agricultural Associations
CPRE	Council for the Protection of Rural England
CPS	Countryside Premium Scheme
CRP	Conservation Reserve Program
CSS	Countryside Stewardship Scheme
DBV	Deutsche Bauenverband
DOE	Department of the Environment
EC	European Commission
EH	English Heritage
ELMS	Environmental Land Management Scheme
EPA	Environmental Protection Agency
EQIP	Environmental Quality Incentives Program

ERS	Economic Research Service
ESA	Environmentally Sensitive Area
EU	European Union
FAIR	Federal Agricultural Improvement and Reform (Act)
FAO	Food and Agriculture Organization
FMI	Financial Management Initiative
FoE	Friends of the Earth
GAEPS	General Agri-Environmental Protection Scheme (Finland)
GAK	Gemeinschaftsaufgabe Verbersserung der Agrastruktur und des Küstenschutzes
GAO	General Accounting Office
GDP	gross domestic product
HEL	Highly Erodible Lands
HLCA	Hill Livestock Compensatory Allowance
HNV	high natural value
IEEP	Institute for European Environmental Policy
ITE	Institute of Terrestrial Ecology
LFA	Less Favoured Area
MAFF	Ministry of Agriculture, Fisheries and Food
MARR	maximum acceptable rental rate
NAA	Nitrate Advisory Area
NACD	National Association of Conservation Districts
NAFTA	North American Free Trade Agreement
NCC	Nature Conservancy Council
NFO	National Farmers' Organization
NFU	National Farmers' Union
NGO	Non-Governmental Organization
NNR	National Nature Reserve
NRI	National Resources Inventory
NSA	Nitrate Sensitive Area
NSCGP	Netherlands Scientific Council for Government Policy
NVZ	Nitrate Vulnerable Zone
OECD	Organization for Economic Co-Operation and Development
OEEC	Organization for European Economic Co-Operation
OTA	Office of Technology Assessment
PIK	payment in kind
PSE	Producer Subsidy Equivalent
RCA	Resource Conservation Appraisal
REPS	Rural Environment Protection Scheme (Ireland)
RSPB	Royal Society for the Protection of Birds
SCS	Soil and Water Conservation Service
SEA	Single European Act
SSSI	Site of Special Scientific Interest
SWCS	Soil and Water Conservation Society
TC	Tir Cymen
UAA	Utilized Agricultural Area
UK	United Kingdom

URAA	Uruguay Round Agriculture Agreement
US	United States
USDA	United States Department of Agriculture
USLE	Universal Soil Loss Equation
WCA	Wildlife and Countryside Act
WHO	World Health Organization
WRP	Wetlands Reserve Program
WTO	World Trade Organization
WWF	World Wide Fund for Nature/World Wildlife Fund

Foreword

Environmental policy for agriculture in Europe and the United States has long been an enigma. While other industries face controls to reduce land, water and air degradation, farms largely have been exempt. Certainly, one can find regulations, such as for pesticides. But historically, agri-environmental policy has relied mostly on voluntary, untargeted educational, technical and financial assistance programmes to persuade farmers to reduce the environmental impact of their activities. This traditional approach is coming under increased scrutiny on both sides of the Atlantic as serious agri-environmental problems persist despite spending billions. Non-point water pollution from cropland in the US and deteriorating European landscapes illustrate the growing tension. Promising experiments in agri-environmental policy reform have emerged during the last decade, but the pace has been slow.

In *Against the Grain* Clive Potter guides the reader through a wide literature to understand agri-environmental policies better in the UK, the European Union (EU) and the US. A fascinating analysis illuminates the different political cultures, policy traditions, and institutional procedures that have shaped the evolution of policy in each region. It is clear that agri-environmental programmes historically have been derivatives of general agricultural policy, but less and less so of late. In the wake of dramatic recent agricultural policy reforms in the US and Europe that lessen subsidies for production and trade two fundamental questions are framed by this insightful review. Will agricultural production interests now shift their attention to environmental programmes as a basis for maintaining public funding of agriculture? If so, what kinds of environmental programmes for farming will emerge?

Answers to the questions hold large import for environmental quality throughout the countryside. It is difficult to overstate the sweep and magnitude of recent reforms in production and trade policies that have intervened in farming since the Great Depression in the US and after World War II in Europe. A string of domestic and international actions to reduce budget and trade conflicts have set farming on course to respond more to market signals and less to government programmes. Shielded from the full force of the market's invisible hand and foot for generations, farmers are entering uncharted territory. Will the shifts in crop and livestock production caused by agricultural liberalization improve or degrade environmental resources?

Potter reviews the hypothesised environmental pros and cons of agricultural liberalization. Some argue that more market discipline will conserve fertilizer, pesticides and natural resources used to excess under former government programmes. Moreover, the savings from cutting programme payments could be used to purchase more improvements, a 'double dividend' for the environment. Others fear that these benefits will not materialize unless policies place full social values on all environmental resources. Without such policies, greater reliance on market prices may cause more volatile and damaging natural resource use. Increased competition may hasten the shift to large production systems with less biological and landscape diversity. Compounding the fears of degradation, certain programmes to conserve land, water and wildlife resources are being dismantled as agricultural liberalization unfolds.

Against the Grain teases out the salient forces that have pushed the US and EU countries to adopt different agricultural and environmental policies, and thereby helps readers to understand why future policies will likely differ as well. It makes clear why the Uruguay Round Agreement was so tortuously long, and why further movement along the trade liberalization path will not come without strong debate about environmental and other matters. One comes away from reading the book with the distinct impression that the agricultural policy reform die is cast for Europe and the US, albeit with different moulds to fit their differing political and cultural traditions.

The path of agri-environmental policy reform remains much less clear. Any number of alternatives can be envisioned based on combinations of subsidy, regulation, government direction and private initiative. Potter discusses two distinct routes. Along the first, former production support is reallocated to measures for public environmental goods such as landscape protection and wildlife habitat. This 'green recoupling' could enjoy substantial political clout by joining agricultural and environmental lobbying interests. Its efficacy hinges upon the government's ability to target high priority problems, place a 'light' hand on management controls that deliver and sustain environmental performance, and avoid paying for agricultural practices that perform at or below the social norm. The second route reduces government intervention by relying on new technologies to simultaneously achieve environmental objectives and capture growing food export demands. Its political saliency, however, hinges on whether less government intervention will be sufficient to meet robust public sentiment for environmental improvements. Government roles shift to more research and technology development that innovate profitable and environmentally beneficial practices and to delivering education and technical assistance to boost adoption. EU countries appear to favour the first route while US interests are consistent with the second. A mixture of the two could easily materialize in each region. *Against the Grain* comes just in time to help readers to navigate this maze of policy evolution with a keen appreciation of history. At stake is no less than one of the next major developments in all of environmental policy.

David E. Ervin
Director, Policy Studies Program,
The Henry A. Wallace Institute, Washington, DC, USA

Introduction

It was inevitable that agricultural policy would attract the attention of environmentalists eventually. Here, after all, was one of the acknowledged drivers of change in the countryside and, according to some, a principal cause of recent environmental loss and decline. Here too was a mechanism designed to transfer large sums of public money into the hands of precisely those individuals with direct management and ownership control over the landscapes and habitats that conservationists were trying to protect. Public unease about the environmental impact of modern agriculture first surfaced in the 1970s, when new evidence concerning soil erosion in the United States (US) and the increasingly visible impoverishment of British wildlife and rural landscapes triggered what Thomas (1983), with pardonable exaggeration, has described as 'one of the biggest movements of public feeling in history'. To begin with, it was individual farmers, the more egregious perpetrators of what Marion Shoard (1980) memorably but controversially called 'the theft of the countryside', who became the targets of criticism in the United Kingdom (UK) as mild concern turned to public anger at the apparent vandalism of habitat loss and landscape change. In the US, Cloud (quoted in Reichelderfer, 1990) describes how:

> environmentalists (who) had spent a quarter of a century repudiating the notion that the tillers of the soil are the best stewards of the land, water and food supply ... finally had the public nodding their heads and farmers shaking theirs, bewildered by their sudden loss of popularity.

What is most interesting in retrospect is how quickly an argument conducted in these largely personal terms, especially in Britain, became a debate about the farm policies and government institutions which were seen to be driving environmental change, often very different in nature, in these two countries.

1

As Weale (1992) has pointed out, policy reform has to start with recognition of a problem and the emergence of an intellectual consensus about what is causing it. This condition had been met by the end of the decade, with broad agreement amongst lobbyists, academics and even government officials that:

> the problem is not one of ill will and ignorance but of a system which systematically establishes financial inducements to erode the countryside, offers no rewards to offset market failures and increases the penalties on farmers who may want to farm in a way which enhances and enriches the rural environment.
>
> (Cheshire, 1985, p. 15)

American conservationists were by this time beginning to ask themselves 'why, after fifty years of federal soil conservation programmes, do we still have a soil erosion problem?' (Batie, 1984, p. 7) and found a convincing explanation in the farm support policies of the United States Department of Agriculture (USDA). The system in question here was the network of commodity price supports, hectarage reduction programmes and investment incentives that had been in place since the Depression and which, harnessed to a technological revolution in farming, had brought about environmentally destructive shifts in cropping patterns and farming techniques (Reichelderfer, 1990). Similar connections between farming policies and their environmental consequences were being made at about the same time in the UK where the European Community's (*sic*) Common Agricultural Policy (CAP) was rapidly becoming the villain of the piece. In this case it was the more general countryside transformations wrought by modern farming practices which became the focus of public concern and an energetic campaign to initiate environmental reform.

Altogether it was a damning indictment of a set of public policies that had been in place in both countries since the 1930s and which, in their basic design, encoded deep assumptions about farming and its relationship to the land. It had long been assumed that farmers had a special role as private producers of public goods, government support being the just reward for secure supplies of food, management of the countryside and economically viable rural communities (Winters, 1987). When they first began to appear in the 1980s, environmental critiques of farm support policy appeared to strike at the heart of this principle because they questioned farmers' ability, under government influence, to deliver an increasingly highly valued range of public goods like biodiverse habitats and beautiful landscapes. In seeking reform, a new breed of agri-environmentalists had set themselves an ambitious task, challenging not only the operation of public policy but also a set of policy entitlements which, historically, had been fiercely defended by a farm lobby renowned for its political agility and entrenched institutional links. That there should be a story about the environmental reform of farm policy to tell suggests a remarkable victory for the alliance of conservationists which pursued the goal of greener farm policies throughout the 1980s and 1990s. In reality, however, that project is still far from complete and some commentators are beginning to question whether many of the agri-environmental reforms are any more than the small change of agricultural policy changes undertaken for largely budgetary reasons. This is probably too harsh a judgement given the undeniable environmental benefits of measures like the US Conservation Reserve Program (CRP) and the European Union's Agri-Environmental Programme (AEP) which, together, have already fixed substantial quantities of conservation capital in the countrysides concerned. Nevertheless, the debate about the effectiveness and political sustainability of these reforms continues and is likely to intensify in the years ahead. The greater visibility of agri-environmental payments compared with traditional market support means that policymakers will come under increasing pressure to demonstrate their effectiveness in environmental value for money terms. Some critics are already beginning to question the desirability of linking environmental concerns to policies which, under newly emerging international pressures for reform, could anyway be substantially dismantled within the foreseeable future (Bowers, 1995).

These are important issues which deserve to be addressed through an analysis of the process of reform and an assessment of its results. For despite the appearance of many excellent commentaries describing agri-environmental programmes, schemes and initiatives (see, for instance, Vail *et al.*, 1994; Whitby, 1996), there are surprisingly few studies of what might be called the political economy of the environmental reform of farm policy, connecting outcomes to the policy processes and political battles which produced them. As Winters (1987) observes, it is difficult to account for the details of public policy, let alone predict its future development, without examining the process from which it springs. Even thinner on the ground are comparisons between countries, despite an acknowledgement by many commentators (see Petit, 1985; Brooks, 1996) that comparative studies can yield important insights into causes, processes and outcomes. The central argument of this book is that in order to appreciate the significance of the reforms themselves, and to predict where they are going, it is necessary to understand why they occurred and how they were accomplished.

It would be too easy in these terms to explain the sudden appearance of green reforms in North America, Scandinavia and Western Europe during the mid-1980s as a straightforward response to mounting public concern about the environmental impact of modern agriculture, however profoundly felt. This has not prevented some commentators pointing to the emergence of agri-environmental programmes at about the same time in the countries concerned as evidence of precisely that. According to a rather pluralist interpretation of events, policymakers eventually responded to the deep shifts in public perception described above, recruiting environmentalists into an expanded agricultural policy community for the first time and adjusting agricultural support to better reflect the public good. Adopting a more sophisticated position, Cox *et al.* (1985a) saw in the ideas of conservationists a major challenge to the privileged position of the farm lobby, and the corporatist arrangements of the post-war period, as long ago as the early 1980s commenting that 'the significance of the growth of environmental concern for the content of agricultural policies is difficult to exaggerate'. Others (see for instance Jordan *et al.*, 1994) have regarded the progressive incorporation of environmental groups into agricultural policymaking as evidence of the policy machine responding to altered public demands, resulting in a widening of membership of the agricultural policy community. Buttel (1992, p. 2) goes further, seeing in the greening of government policy 'larger currents of social change in the late twentieth century', though he accepts that more contingent factors may explain the achievement of green reforms in practice. For Majone (1989, 1992) in particular, the way participants marshall evidence and deploy ideas to challenge the assumptions of their opponents, is always going to be an important part of any explanation of policy change. Without succumbing to the rationalist fallacy that theories and ideas are alone powerful enough to determine the course of events, Majone maintains that policy reform is just as often brought about by changes in belief and values as by realignments of economic and political interests. According to this model 'participants marshall evidence in support of their proposals ... and make arguments that appeal to the beliefs and values, as well as the interests, of broader constituencies' (Majone, 1989, p. 148). Sometimes they will seek to challenge core assumptions which have traditionally underpinned public policy and legitimated its existence. Occasionally they succeed in levering policy change by bringing about a re-examination of programmatic policy doctrines.

One of the most important of these in the agricultural policy case is the idea that prosperous family farmers make the best environmental stewards. As Chapter 1 will show, it was a key contention of agri-environmentalists that this was not necessarily true and that the attempt to boost farm incomes through 'coupled' systems of agricultural support had contributed significantly to the environmental crisis of modern agriculture. By the early 1980s, the structural connection between farm support and environmental change had been well documented in both the US and, within Western Europe, the UK. At the same time there was evidence of a gradual but cumulative change in the public's attitude to farmers and farming which Swanson (1989), amongst others, interprets as evidence of a shift in the cultural context of agricultural support. Indeed, 'concerns over groundwater pollution (and) dangerous pesticide residues in food ... not only represented signs of a changing cultural context for agricultural policy, they also (anticipate) the introduction of non-farm groups into its formulation' (Swanson, 1989, p. 82). Rather as Paarlberg (1988) has predicted, urban interests were beginning to be substantially involved in the agricultural policymaking process for the first time, bringing with them new environmental and food quality concerns.

All this being said, the arguments of environmental reformers, however well articulated, were never likely to be sufficient to trigger policy change, and to rationalize the events which followed strictly in these terms would be to risk over-determining the process of agri-environmental reform. As Sabatier (1987) argues, in the real world it is necessary to identify the interests behind the ideas and to recognize that policy change, in the short term at least, is usually the product of political bargains struck between groups competing for the attention of policymakers and the resources at their disposal. It is doubtful, for instance, that the early campaigns of agri-environmentalists would have borne much fruit without the first tacit, and later, overt, support of a farm lobby looking to find new ways of legitimizing government subsidies in a period of general policy retrenchment. Chapter 2 explains why the onset of agri-environmental reform in many industrial countries at roughly the same moment is much more likely to have been due to budgetary pressures than environmental ones. The effect of these was to force a reassessment of farm programmes in general, making policymakers in their turn more receptive to the ideas then being put forward by conservation groups in both the US and in Western Europe. As Moyer and Josling (1990) have observed, the reform of a policy as culturally entrenched as that for agriculture rarely takes place unless precipitated by external pressure, usually of a financial sort. In their words:

> what makes a budgetary crisis so threatening to both agricultural bureaucracy and agricultural legislative actors is that it turns the policy process into a zero-sum game where agriculture's demand for new resources is seen as coming at the expense of other programmes. No other crisis is likely to create the same kind of outside interest.
>
> (Moyer and Josling, 1990, p. 10)

By mid-decade, agricultural overproduction had put the farm support systems of both the US and the European Union (EU) under severe financial strain and policymakers were looking for ways of limiting budget exposure. For its part, the farm lobby was looking for new allies in order to defend the principle of government support to the agricultural sector. In the US particularly, the passage of the CRP under the 1984 Farm Bill was greatly eased by the perception that it would tackle overproduction and offer farmers another source of public subsidy at a time of impending policy retrenchment. More obliquely, the EU's AEP of the early 1990s was deployed as a mechanism for farm income support as well as environmental protection in a period when farm ministers were making their first moves to limit price guarantees under the CAP. In both cases, the outcome was a substantial investment of public funds in programmes to subsidize environmental management on farms.

The middle section of the book compares the experiences of the US, the UK, and the EU as a whole, in drawing up and implementing these programmes to ensure that agri-environmental reform actually came about. Differences in political culture, policy traditions and institutional procedures mean that agri-environmental policy development has proceeded at different rates and followed diverging trajectories in the countries concerned. If it was a coincidence of budgetary pressures with environmental concern that precipitated reform, the nature of the policy changes achieved has been heavily influenced by the national agricultural policy context in which they have taken place and what Weale (1992) calls the standard operating procedures of bureaucratic politics. In the case of the US, the practice of legislating for agriculture through 5-yearly omnibus farm bills created an opportunity agri-environmentalists were quick to exploit. The result was the rapid appearance of an ambitious Conservation Title under the 1985 Farm Act which built large ideas like 'conservation compliance', and the CRP, into federal policy in order to address a class of environmental problems that were largely defined in scientific and normative terms. As Chapter 3 explains, however, the Title was rather less revolutionary than it seemed, drawing directly on 1930s New Deal policy traditions and initially defining the problem (soil erosion) in ways which often privileged farming interests at the expense of environmental ones. In describing the subsequent implementation of the Title, this chapter shows how pressure was applied to improve its environmental performance once it became clear that the programme would be at risk if it could not be shown to be good environmental value for money. Having invented a comprehensive policy, lobbyists and policymakers, it would seem, have been preoccupied with finding better ways to justify its existence.

The contrast with the UK's policy style could not be more stark, for here agri-environmental policy has been a much more incremental affair, the product of a slower moving policy process. Parastatal organizations unique to the UK like the Countryside Commission (CC) have played a key role as policy entrepreneurs and risk takers and there has been a greater emphasis than in the US (or indeed, throughout the rest of the EU) on the need for experimentation from pilot schemes. Chapter 4 locates the origins of UK agri-environmental policy in a series of land use conflicts which challenged traditional site-based approaches to landscape protection and habitat management. It describes how the Ministry of Agriculture, Fisheries and Food (MAFF) and its sister agriculture departments in Scotland and Wales set up a series of measures designed to promote the very British idea of countryside management and discusses the continuing debate here about environmental reform of the CAP. By drawing up schemes which paid farmers to 'produce countryside', the UK was also piloting an approach which would eventually be taken up throughout the EU. Chapter 5 analyses the further evolution in thinking which accompanied the development of an AEP for the EU as a whole and the attempts by the European Commission (EC) to reconcile often radically different national conceptions of what agri-environmental measures should be designed to achieve. As Majone points out, 'issue transformation' is a typical feature of incremental policy change, the creation of stable coalitions around particular conceptualizations of the problem being an essential prerequisite for policy reform. Here, growing concern with the specifically public health effects of agricultural pollution merged with a longer established concern about landscape change and agricultural decline to make a strong case for agri-environmental reform under the CAP. Meanwhile, by inventing the idea of 'extensification', the EC was able to justify payments to farmers on a much wider front than would be possible under the UK model alone. The result was a more open-ended and permissive policy dedicated to the prevention and control of agricultural pollution as well as the management of the countryside.

By the mid-1990s it was possible to look back on over a decade of solid policy achievement, much of it in the early stages taking place against a background of agricultural overproduction and budgetary constraint. While it was often too early to judge the precise environmental impact of the various schemes now on offer to farmers, there has been wide agreement that most agri-environmental programmes had improved the incentives for environmental management and conservation on farms. Exchanges of information and transfers of best practice between countries though were rare, important differences in the nature and scale of the environmental impact of agriculture in North America and Western Europe intersecting with policy traditions and given institutional structures to create many variations on the common theme of green subsidization. With international agricultural policy reform under the aegis of the World Trade Organization (WTO) now in prospect, however, commentators are predicting much more convergence towards a common set of principles for agri-environmental policy design. This is the theme taken up in the penultimate chapter of the book. Signatories to the Uruguay Round Agriculture Agreement (URAA) of 1993 have committed themselves for the first time to a common approach in tackling international agricultural policy reform, the immediate effect of the Agreement being to accelerate the 'decoupling' of farm support that has been in process in the US since the early 1990s and to establish this as a model which the EU will increasingly have to follow. As Harvey (1995, p. 210) puts it, 'Farm prosperity is now increasingly understood to be unsustainable through market price support, and only achievable through an internationally competitive industry'. Consequently, major changes to the structure of farm support and the way it is publicly justified and defended are in prospect, sweeping agri-environmental policy into the larger debate about the liberalization of agricultural trade. Where government support to agriculture continues, it is likely to be justified in more direct 'purchase-provider' terms than ever before, farmers being contracted by the state to produce the public environmental goods an increasingly wealthy society demands. Arguably, this injects a new dynamic into the debate as subsidies to farmers are argued for in terms of individual merit rather than collective desert. Chapter 6 analyses what this is likely to mean for agri-environmental strategies that, until now, have been predicated on a continued reform of agricultural policy, not its possible abolition. It assesses how far the scaling down of market price support and the 'decoupling' of public subsidies is likely to benefit the environment and compares the American and European case for retaining some form of government support to agriculture on environmental grounds. Finally, in Chapter 7, the threads of the book are drawn together to present a comparison of the origins, evolution and likely future development of agri-environmental policy in the US and the EU. This chapter reflects on what has been achieved through the agri-environmental reforms of the 1980s and 1990s and seeks to identify what countries can learn from each other in subsidizing farmers so that they can better fulfil their stewardship obligations. It points to the evolutionary potential of the policy experiments already in position and explains why, for the foreseeable future, agri-environmental policy is set to remain a permanent feature of the rural policy scene.

1 Engines of Destruction?

Agricultural policy has long been regarded as one of the deep engines of rural change in industrial countries. As a social institution, though, it is a relatively recent invention and the exact nature of its environmental impact and influence is even now far from clear. Governments first began to intervene systematically to support the incomes of farmers during the 1920s and 1930s when economic depression pushed farm prices to levels low enough to imperil the livelihoods of large sections of the farming community. Paarlberg (1989) points out that agricultural policies have usually been brought into existence by governments responding to an economic emergency in order to protect the interests of those perceived to be vulnerable members of society. Without the shock of the Depression or World War II it is unlikely that American or European governments, respectively, would have intervened so deeply in the agricultural economy when they did. The historical record shows governments having surprising freedom of action, initially at least, in deciding the level and form of support (Benedict, 1953; Tracy, 1985). Observing the rapid invention of federal commodity support in the US in the early 1930s, Finegold (1982, p. 23) comments that:

> the pluralist conception, accurate in many contexts, is that groups represent individuals and influence the state. After Roosevelt's election, however, it was the President-elect and his advisors who were able to use the ties developed during the election campaign with the farm organizations to 'impose' their preference for a policy of production control upon the organisations concerned.

The Great Depression had cut the prices farmers were receiving for their products by over 50% in the 3 catastrophic years between 1929 and 1932 and one of Roosevelt's first acts as President in 1933 was to offer government assistance on a broad front (Benedict, 1953; Finegold, 1982). The Agricultural

Adjustment Act of 1933, one of the cornerstones of the New Deal, gave the Secretary of Agriculture sweeping powers to increase farm prices and so boost farm incomes by removing commodities from the market and by controlling output from domestic farms through a system of allotments specifying the areas to be planted to crops. From an analysis of the problem which linked persistently depressed farm incomes to a chronic tendency towards over-production followed a policy designed to keep the massive productive potential of American agriculture at bay through hectarage controls. The USDA was transformed from a small-budget organization mainly concerned with re-search and education into a government agency responsible for injecting large amounts of public money into rural America. As Wilson (1977) comments, one of its legitimizing assumptions was that keeping large numbers of family farms on the land was a socially and culturally desirable thing to do. The roots of this policy had been nourished by a form of agrarian fundamentalism which saw farming as a way of life and farmers as Jeffersonian 'builders of the nation' and the stewards of its natural resources (Bonnen and Browne, 1989). While the resulting policy of protection could be rationalized in socio-economic terms when over a quarter of Americans still lived on the farm and there was mass unemployment in the larger economy, it would very soon be operating against the grain of a technological revolution in farming which made it possible to move as many as one-third of farmers into other industries (Benedict, 1953; Tweeten, 1979). It would also be challenged by free marketeers who could point to the comparative advantage of American agriculture on world markets and thus to the benefits of an export-led growth in agricultural output pro-duced on fewer, larger and more efficient farms (Brandow, 1977).

European policymakers approached agricultural policy from a different direction, though ultimately their justification and style of reasoning was much the same. With an agricultural industry that is much less productive and more fragmented than America's (the average size of American farms in 1940 was 67 ha compared with just 10 ha in France), countries like France and Germany already had a tradition of agricultural protectionism designed to protect their large farm populations from the vagaries of world markets. In Germany, particularly, the desirability of maintaining a large agricultural population was an assumption so deeply rooted as to be unquestionable and since the 1890s governments had implemented a system of import levies on grains in order to protect farmers from the worst effects of foreign competition. Economic depression drew European governments further into agriculture so that by the mid-1930s price support in some form had been established in The Netherlands, Denmark, Norway, Sweden and France, as well as Germany (Tracy, 1989). Fennell (1979, p. 2) quotes a Food and Agriculture Organ-ization (FAO) study, which observes the first substantial moves by Western European governments at this time to 'shield their domestic producers against the worst effect of the depression. In so doing, they not only reduced their dependence on the outside world, but also restricted inter-country specialization within Europe'. Even in the UK, where a *laissez-faire* tradition dating from the repeal of the Corn Laws in 1849 had brought about the restructuring of production so conspicuously lacking in most other Western European countries, price support was being offered on a broad front by 1932. This took the form of deficiency payments which bridged the difference between market prices and target prices set by government. Marketing schemes and import restrictions were rapidly added to the apparatus of farm support. As Smith (1990, p. 70) observes, 'A nation committed to free trade was suddenly prepared to protect agriculture', and to do so on an ambitious scale. By the mid-1950s, British farmers were enjoying levels of government support that bore comparison with any of their continental counterparts (McCrone, 1961).

Working with the Grain

World War II confirmed agricultural support as a permanent feature of the policy scene in both the US and Europe. It is at this point that agricultural interest groups emerged as important players in the policy process for the first time, exerting pressure after the war to retain war-time price guarantees. As Paarlberg (1989) argues, organizing to defend policy benefits already in place proved to be a relatively easy task. Once support has been instituted, agricultural interest groups mobilize quickly to defend their policy entitlements, entrenching farm programmes institutionally and politically and acting like a ratchet to prevent retrenchment. Public Choice theory and the New Political Economy of agricultural policy explains their remarkable ability to do so in terms of the logic of collective action and the 'rational ignorance' of the taxpayers and consumers who ultimately bear the cost. The former operates because farmers discover that, with economic development, they not only have an increasing economic incentive to organize collectively but they are better at it. The incentive derives from the operation of Engel's Law and the effects of technical progress. The former says that a smaller share of the average consumer's household budget is spent on food with every *per capita* increase in household income. At the same time, unstoppable productivity growth in agriculture, due to endogenous technical progress, means an expansion in supply at the very moment demand stagnates. Without government intervention, farming is thus a declining industry in which it becomes increasingly difficult to earn a decent living. Even with government support, the farming population in the US and EU has declined significantly since the 1930s and with smaller group size individual farmers become more willing contributors to political action. A formal rationalization of this is offered by Olson (1965), who predicts that small groups are less prone to free riding and thus more successful in retaining the collective loyalty of their members (though a more empirical and substantive explanation of the success of farm groups would also need to point to their steady incorporation within the policy machine and its standard operating procedures and their success in exploiting cultural myths about farmers to legitimate their demands – see, for instance, Smith (1990) and Browne (1988)). Taxpayers and consumers, on the other hand, are said to be 'rationally ignorant' of the costs of agricultural support because the costs to them as individuals of becoming better informed and doing something about it outweigh any benefits. In addition, as Brooks (1996) points out, with declining farmer numbers, a greater *per capita* subsidy can be provided at lower *per capita* costs to consumers and taxpayers. The result, according to this 'cosmopolitan' model, is the near universally high levels of government support enjoyed by farmers in industrial countries over the last four decades.

The historical record shows that American farmers actually enjoyed something of a post-war boom as prices and incomes rose steadily in response to world scarcity and farmers were exhorted to feed the destitute of Europe. At the same time, government war-time guarantees meant that price support was set at 90% of parity until 1948 (Cochrane and Runge, 1992). The farm lobby's immediate concern after the war was thus to preserve this level of support. A brief struggle between economic conservatives and New Deal technocrats towards the end of the decade resolved itself in favour of continued government intervention, despite the strong intellectual case for an open market policy. Under the Brannan Plan of 1948 policymakers were offered an alternative way of supporting the incomes of farmers through degressive direct payments which would provide a safety net for the smallest and most marginal farms while freeing the large productive farms to produce for world markets (Christenson, 1959; Tweeten, 1979). By this time it was becoming clear that price support was an inefficient and self-defeating way to support farm incomes because it gave an incentive to the most efficient farmers to overproduce and necessitated expensive offsetting production control programmes. Agriculture Secretary Brannan's argument was that only a 'decoupled' system of government payments could be financially and politically sustainable, and proposed the abolition of price support in favour of a system of direct income payments expressly designed to support family farms – the standard formulation of agricultural economists ever since. As Wilson (1977, p. 61) puts it:

> the Brannan Plan would have avoided many of the faults of the existing farm laws. By substituting government payments for artificially high market prices, it would have encouraged consumption. Above all, ... the Plan would have allowed American commodities to compete freely on world markets.

In rejecting the Plan, policymakers turned their faces against a radical overhaul of agricultural policy and the New Deal reasoning on which it was based, and set the scene for the policy conflicts of the next 30 years. These revolved around the problem of overproduction and the best way to deal with it.

The dilemma for American Secretaries of Agriculture has always been the same: how to keep surpluses under control without subjecting the politically powerful farming community to excessive economic pain. With the mechanism of 'cross compliance' – giving support through commodity payments on the one hand but threatening to take it away with the other if farmers did not comply with the requirement to set some cropland aside – they thought they had found a reasonable compromise. The first significant experiment with large-scale land diversion came in 1956–1958 when over 8.5 Mha of cropland were taken out of production under the Soil Bank programme. Farmers could choose between annual contracts for withdrawing part of their arable area from production or for the long-term retirement of all cropland from production (the conservation reserve option). Despite evidence that the conservation reserve succeeded in withdrawing sub-marginal land from production (Brandow, 1977), the programme failed to stem the expansion in production. This was because farmers enrolled their least productive land first and then continued to intensify on the land which remained in production. They also shifted production into alternative crops not covered by the set-aside requirement (Brandow, 1977). Undeterred, USDA was back with a voluntary hectarage reduction programme in the early 1960s and again later in the decade (see Fig. 1.1). By this time the Kennedy administration was also promoting the export of surplus production under Public Law 480 and other

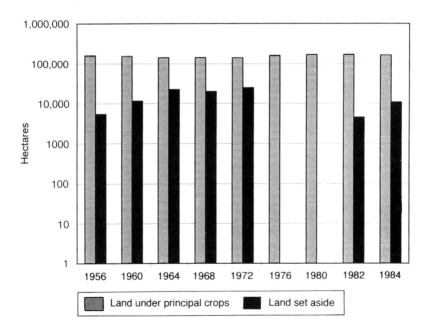

Fig. 1.1. Land retirement under USDA programmes, 1955–1985. Note: a log scale has been used. Source: USDA (1985).

indirect subsidies and under the 1973 Agriculture Act the Acreage Reduction Program (ARP) was dropped and farmers were encouraged to 'plant fence-row to fence-row' in order to exploit an expansion in world market trade. For a period the commodity programmes were eclipsed by soaring world market prices as farmers looked increasingly to the export market for the best prices. At this stage, US farm exports were worth over a billion dollars per year, PL 480 effectively expanding foreign demand for US farm products (Cochrane and Runge, 1992). This bubble burst in 1980 with the US embargo on grain exports to the USSR and by the early 1980s the system of deficiency payments linked to hectarage reduction that had last been rebalanced under the 1973 Act was coming back into play. Indeed, due to an underestimation of the 1982 harvest and thus a programme of hectarage control that was too small to achieve market balance in that year, the most conservative administration of the post-war period found itself intervening in the agricultural sector on a massive scale during 1983 when it set up the Payment in Kind (PIK) programme, paying for land diversion 'in kind' with stocks of agricultural commodities already owned by USDA. Altogether, 31 Mha were idled under PIK, an area equivalent to the entire planted hectarage of Western Europe in that year (Paarlberg, 1989). Coinciding with the 1983 drought, the effect was a sharp fall in output and a significant loss of world market share. The scene was being prepared for what commentators would soon be calling the farm crisis of the 1980s.

In Europe, overproduction was still a remote possibility in 1945. Agriculture throughout the continent had been ravaged by the campaigns of 1944 and 1945 and, in Germany particularly, productive capacity had been seriously reduced by Allied action (Tracy, 1989). Galvanized by the need to increase output and aided by the Marshall Plan, governments reinforced existing price guarantees, instigated income aids and subsidized capital investment throughout the industry. Under the UK's 1947 Agriculture Act, the government undertook to purchase the entire domestic output of grains, potatoes, sugar beet and fatstock at guaranteed prices and instigated a system of deficiency payments to bridge any gap between market and target prices. Deficiency payments were not feasible elsewhere, given the much larger farming populations with which policymakers had to deal, but price support effected through import levies and intervention on domestic markets was, and by the early 1950s farmers throughout Western Europe were sheltering behind high tariff barriers as policy took a decidedly protectionist turn. In Germany especially, agriculture was exempted from the free-market disciplines which were in the course of being applied to other sectors, the Deutsche Bauenverband (DBV) successfully pressing the case for special support to boost farm incomes and retain the family farm (Gardner, 1996). French attitudes were more complex, a similar commitment to defending the residual peasantry competing with a growing mercantilist concern to improve agricultural productivity and export potential (Neville-Rolfe, 1984). The very rapid expansion of agricultural output here, and in other Western European countries, in this immediate post-war period suggests that, like their American counterparts, policymakers were harnessing a technological revolution in farming. By 1949 output in the area covered by the Organization for European Economic Co-Operation (OEEC) exceeded the pre-war level and by 1950 was 50% higher (Tracy, 1989). During the next three decades agricultural output in the six founder states of the EU would expand by over 3% per year on average (CEC, 1995). Unlike the US, however, this was expansion from a very low base. European agriculture at mid-century was still largely peasant based and the jump in production was being achieved through the much more wholesale replacement of 'one mode of production dependent on horsepower and man-power to one based on tractors, chemicals and oil' (Duchêne *et al.*, 1985, p. 21). A reflection of this is the fact that while, between the two World Wars, the farming workforce fell by only 10%, land in cultivation remained stable and output increased only marginally; after 1950 manpower was cut by two-thirds, the land area by 10% but output multiplied by two-thirds (Duchêne *et al.*, 1985).

The CAP came into existence just as this revolution in outlook and tech-nique was getting into its stride. Given the economic importance of agriculture in the six founder member states of the European Community (France, West Germany, Belgium, The Netherlands, Luxembourg and Italy) it was inevitable that the common organization of agricultural markets would be part of the customs union that was formed under the Treaty of Rome in 1958. In the mid-1950s those employed in or connected with agriculture accounted for a quarter of the civilian population in these six founder states and 10% of their combined gross domestic product (GDP; Wilson, 1977). Moreover, it was part of the initial political bargain behind the Treaty of Rome that France, The Netherlands and Italy would open their borders to German industrial goods only in return for being given privileged access to that country's agricultural market (Tracy, 1989). The Treaty lays down as a central aim of the CAP that of ensuring 'a reasonable standard of living' for the agricultural community, going on to suggest that this would be achieved through increased produc-tivity and increasing the individual earnings of those engaged in agriculture. In practice, this connection, implying widespread structural change, and the weeding out of small, uneconomic farms, would never properly be made and from the beginning policymakers attempted to maintain the incomes of a large population of marginal producers by offering high price guarantees to all farmers. Fennell (1985, p. 260) comments that:

> the reality of the CAP is that it is concerned, above all, with the pursuit of production rather than the promotion of productivity. Such an outcome was inevitable given the over-riding importance of a flat-rate price support system as the main instrument of policy, the impact of which is to ensure that the greater the number of units a farmer produces the greater the amount of support received.

In fact, at a very early stage it was decided to carry over the existing protectionist arrangements prevailing in the founder member states based on
variable import levies and government intervention to maintain internal
market prices, and during its first year of operation the CAP pushed farm prices
well above world market levels. For a long time, policymakers did not have to
deal with the consequences of the inevitable further expansion of agricultural
output which followed. Indeed, improving rates of self-sufficiency was one
of the original aims of the policy, while the accession of the UK in 1973, at
that stage a net importer of agricultural products, gave further room for
manoeuvre. The effect, as Duchêne *et al.* (1985) observe, was to create slack in
European import markets for food exports from other member states with
preferential access.

In 1968, however, Agriculture Commissioner Sicco Mansholt decided to
point out that the Emperor would soon not have any clothes. Mansholt's
central argument was that the only long-term solution to low farm incomes
was fundamental structural change. Blanket price support was no answer
because it subsidizes larger and more efficient farmers disproportionately and
leads to the creation of surpluses, leaving untouched those producers on the
margins who lack the capital and expertise needed to become viable businesses. Moreover, a high price policy was ultimately self-defeating in a period
of rapid technical change, because eventually a point would be reached when
the budgetary strain of having to remove huge food surpluses from the internal
market in order to support prices would necessitate a drastic cut in price
support, wiping out the very family farms the policy was supposed to protect.
His words had a prophetic ring:

> The Community is now saddled, for many commodities, with surpluses some of
> which have no prospect of outlets on saturated world markets. Where outlets do
> exist, surpluses depress prices so that they can only be sold at high cost to the
> Community's budget ... It is therefore vital to adopt another policy for farm
> prices.
>
> (CEC, 1969, p. 34)

The resulting Mansholt Plan, presented to the Agriculture Council in
December 1968, proposed a radical programme in which agriculture departments would buy out these small producers and reallocate the land given up to
larger units. People leaving agriculture would either take early retirement or
be encouraged to retrain for another occupation. The Plan envisaged a
relatively small reduction of around 7% in the area under cultivation but
a very large reduction in the number of people employed on it – 5 million, or
about half of the existing workforce (Nevin, 1990). This was Europe's equivalent of the Brannan Plan and, as in the US, its swift rejection by farm groups
and farm ministers (see Rosenthal, 1975), fearful of the long-term structural
implications of the Plan, confirmed traditional approaches and put policy on
the trajectory it would follow for the next 30 years.

Those in charge were rejecting the option of a low price policy which could restructure the industry and raise the living standards of a smaller community of competitive farmers, in favour of a farm survival policy which would attempt to improve the incomes of a large number of farmers through modernization and market protection (Potter, 1990). The underlying rationale for this strategy, to the extent that it was ever actually articulated at this stage in public debate, was that agricultural populations had to be maintained in order to preserve the social well-being of the countryside and to prevent depopulation and desertification. According to this 'Green Europe' view of rural development, dating back to the Stresa conference setting up the CAP, a US-style countryside of 'large reserves of land and few farmers' was inappropriate for the EU because of the value attached to a peopled countryside and a 'landscape with figures'. It was thus legitimate to use public money to support the incomes of farmers. Price support now rose steadily following the institution of the 'objective method' to fix annual price awards. This entailed calculating the percentage increase in guaranteed prices necessary to keep the incomes of 'modern' farmers in line with non-farming incomes, a formula which was soon being exploited by the Committee of Professional Agricultural Associations (COPA) to procure generous price awards (Phillips, 1990). Later, after an unprecedentedly generous award in 1974, the basis of the calculation changed, though it was not until the advent of the system of guarantee thresholds in 1982 that the policy of offering farmers unlimited price guarantees in an effort to increase the incomes of the most marginal producers finally came to an end. Under the Structures Regulations of 1972, meanwhile, member states were encouraged to offer generous investment aids to farmers capable of bringing the income-earning capacity of their businesses up to the regional industrial average. The main thrust of this modernization policy, as Fennell (1985) points out, was to encourage farmers to invest in new machinery, buildings and equipment in the hope that this would solve the farm income problem of the existing community of farmers. In a half-bow at Mansholt, member states were also empowered to offer early retirement schemes to elderly farmers in order to speed up the adjustment process (though payment rates would be set too low to have more than a marginal impact here). Indeed, in 1975 the Community moved further away from a Mansholtian approach by agreeing the Less Favoured Areas Directive. This permitted the use of EC funds to keep marginal farmers on the land by designating disadvantaged regions and offering annual headage payments and other forms of assistance to the farmers concerned. A decade later over 48% of the Community's land base had been so designated and by 1995 56% of the Utilized Agricultural Area (UAA) of the EU was designated under the Directive (see Table 1.1). Annual expenditure on these supports in that year totalled 1.4 billion ECU (CEC, 1995). Policymakers were now in pursuit of two apparently contradictory policy goals: keeping large numbers of farmers on the land while at the same time attempting to bring about a significant improvement in average farm incomes. It would not be long before

Table 1.1. Less Favoured Areas within the EU. Source: CEC (1996c).

	Land area in thousands of hectares					
	Mountain areas[a]	Other less favoured areas[b]	Less favoured areas with specific handicaps[c]	Less favoured areas	Country total	Less favoured area as % of total UAA
Austria	2047	208	164	2419	3524	69
Belgium	–	273	–	273	1357	20
Denmark	–	–	–	0	2770	0
Finland	1407	536	220	2164	2549	85
France	5284	7809	804	13,897	30,011	46
Germany	336	7987	199	8522	17,012	50
Greece	3914	964	402	5280	6408	82
Ireland	–	3456	12	3468	4892	71
Italy	5218	3405	218	8841	16,496	54
Luxembourg	–	122	3	125	127	98
The Netherlands	–	–	111	111	2011	6
Portugal	1227	2056	150	3433	3998	86
Spain	7503	11,343	700	19,546	26,330	74
Sweden	526	1011	333	1869	3634	51
United Kingdom	–	8341	1	8342	18,658	45
Total EU	27,462	47,511	3317	78,290	139,777	56

[a] Designated under Article 3 (3).
[b] Designated under Article 3 (4).
[c] Designated under Article 3 (5).

the irreconcilability of these goals became all too clear. To quote Gardner (1996, p. 18) 'the stage was [now] set for the surplus-producing, budget-busting farm production expansion of the 1980s'.

Brakes and Accelerators

Indeed, if there is a common thread running through the subsequent history of agricultural policy in industrial countries, it is the way public policy has interacted with technological change to create a situation (agricultural over-production) which requires a further policy response. As Cochrane and Runge (1992) put it, technological change has driven farm policy even as it has been driven by it, forcing policymakers, first in the US and more recently in the EU, to acknowledge the chronic tendency of the agricultural industry towards over-supply. Grasping this fact is the key to understanding how agricultural support has contributed to environmental change throughout rural Europe and America over the last 30 years. The story begins early in the century in the US when pressure of demand and a drive to reduce production costs strengthened the incentive for innovation, and thus research. Buckwell (1989) observes that technical progress is never fully explicable in economic terms and to a degree is 'manna from heaven', the spontaneous result of the expansion of knowledge. In the American case, however, a documented increase in the cost of land relative to labour over this period provided a spur for the mechanization of agriculture which took place during the 1920s, and for the biological–chemical revolution in farming techniques which followed swiftly on its heels. Investments in research and development effected through the system of Land Grant colleges established in the 1890s also now began to pay off as innovations like hybrid corn were rapidly adopted by large numbers of farmers (Cochrane, 1980). The County Extension Service, set up in the 1920s, very effectively lowered the transaction costs of adoption, communicating back to USDA and the Land Grant colleges information about farmer resistance to particular innovations and suggesting how they could be overcome (Rausser and Foster, 1989).

According to the treadmill theory of technological change, however, none of this was without structural consequences (Cochrane, 1958; Levins and Cochrane, 1996). The early adopters of the new technologies enjoy a reduction in production costs which gives them an advantage over laggards. As prices and net returns begin to fall in the face of rising output and sluggish consumer demand, average farmers are forced to adopt the technology if they are to survive. The laggardly farmers who fail to do so are lost in the price–cost squeeze, giving up their land and resources to the smaller number of productive farms which remain. In reality, the government intervened before this outcome was achieved, acting to support prices and incomes as output increased. The effect was to maintain the incomes of the average farmer but to boost even further the profits of early adopters. As Heady (1984) remarks, the bounty in farm incomes at this time did not come from exploding international markets but from Treasury-supported prices. An inflation of land values followed during the late 1940s as a large number of expanding farmers competed to obtain more land in order to take advantage of government-guaranteed prices but discovered that these same government programmes were reducing opportunities for amalgamation by keeping marginal farmers on the land. In Cochrane's terms, a technological treadmill was replaced by a land market treadmill as, for the next 40 years, farmers bid up land prices on the basis of expectations about a continuation of commodity support: with 1936–1940 values equal to 100, land prices had risen to 196 by 1968–1972 and to 342 by 1981–1985 (USDA, 1989). The immediate result was a further encouragement to land-saving technologies as American agriculture began to shift decisively towards a land-saving mode. A general intensification of production was soon under way, with fertilizer use doubling each decade between 1940 and 1970 and farm chemical use more than doubling over the same period (though, as Crosson (1991) points out, after 1970 fertilizer use levelled off somewhat as input costs rose relative to crop prices and diminishing returns to further increases in fertilizer use began to set in). At the same time land was drained, reclaimed and brought into production, the area of drained agricultural land doubling to 42.5 Mha by 1980. The irrigated area more than doubled between 1950 and 1977, exceeding 20.2 Mha by the end of the decade (Crosson, 1991). Government land retirement programmes accelerated these processes of intensification by tightening still further the land constraint on farmers. The effect, in technological terms, was to give a spur to the development of land (and labour)-saving technologies, Offutt and Shoemaker (1988, p. 2) concluding from their analysis of time-series data that 'by artificially restricting the supply of land, [ARPs] have caused technological change in farming to be even more biased towards land-saving innovation than would otherwise be the case'.

The environmental consequences of this simultaneous braking and acceleration of agricultural change did not become fully apparent until the mid-1980s when a study commissioned by USDA (Reichelderfer, 1985) documented the contribution made by federal commodity programmes to soil erosion on US cropland. Over the preceding 25 years, it was claimed, federal commodity programmes had shed any soil conservation associations they might once have had. The conservation conditions attached to set aside were minimal, while the incentive to intensify production on cropped land was intense. Ironically, it was the consequences of government taking its foot off the brake which first drew the attention of modern environmentalists to what Bennett, four decades earlier, had called 'the national menace' of soil erosion (Bennett, 1931). By mothballing hectarage reduction programmes and exhorting farmers to plant 'fence-row to fence-row', the government brought about an oceanic change in land use during the mid-1970s which swept away terraces, shelterbelts and other conservation practices installed under the New Deal programmes to keep soil erosion under control. Some 13 Mha were converted to crops between 1972 and 1975, many of them very vulnerable to erosion, and by 1981 the area of cropland was greater than at any time in the nation's history, 37% of which was devoted to production for export (Batie, 1984). By late 1973 USDA was receiving reports of farmers throughout the Great Plains 'ploughing up grassed waterways, shallow hilltops and steep slopes ... and tearing out windbreaks that took many years to establish'. From the southern Great Plains there were 'stories of speculators breaking new ground and preparing to plant cotton on thousands of native rangeland [hectares] that had never been used for crops before' (Helms, 1990, p. 12). This period also saw a steep decline in conservation investment on farms. Thigpen (1994) quotes USDA sources which show a fall in gross annual conservation investment measured in 1977 dollars from $911 million in 1961–1965 to $640 million in 1966–1970 and only $445 million in the 1971–1975 period. The impact on soil erosion was soon apparent. According to the National Resources Inventory (NRI) that was published in 1977,

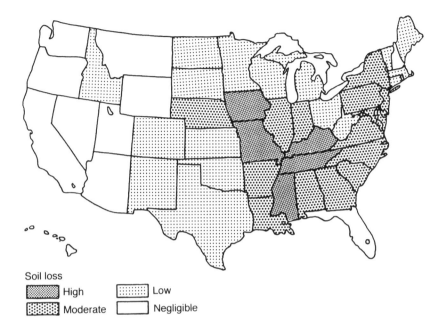

Soil loss
| ▓ High | ░ Low |
| ▒ Moderate | ☐ Negligible |

Fig. 1.2. Incidence of soil erosion in the US, 1977. Source: Batie (1984).

1.8 billion tonnes of soil were being lost annually because of sheet and rill erosion and a further 885 million due to erosion by wind (USDA, 1981). Five years later, the 1982 NRI put an even larger figure on erosion due to the effects of water. The national average loss of soil on cropland due to water erosion, based on the Universal Soil Loss Equation (USLE), was estimated at 1.9 t ha^{-1}. There were wide variations by state, however (see Fig. 1.2). Tennessee headed this list, with an estimated loss of 5.9 t ha^{-1} in 1977. Wind erosion losses in ten Great Plains states averaged 2.1 t ha^{-1} in this year (Batie, 1984). In response to mounting public concern, USDA undertook the Resource Conservation Appraisal (RCA) in 1981 which gave lengthy treatment to soil erosion and its impact on both productivity and water quality.

By this time the longer term impact of the commodity programmes themselves on the pattern and intensity of resource use was becoming clear. While the broad geographical distribution of production has changed very little (the Corn Belt, Northern Plains, Mountain and Pacific states had 55% of the nation's cropland in 1940 and still accounted for 65% in 1987: Crosson, 1991), the mix of crops grown on individual farms has altered dramatically. A massive policy-induced shift away from relatively soil conserving crops like wheat, oats and barley in favour of more heavily subsidized, exportable and erosive row crops like cotton, soybeans and corn, meant that more vulnerable cropland was exposed to the forces of erosion than ever before. As Reichelderfer (1990, pp. 204–205) points out, price support and technological change have together tailored the evolution of crop production towards monocultural systems that are inherently chemical dependent and destructive of the soil,

> the upshot [being] that the US Government is effectively subsidizing the production of precisely those crops most prone, by virtue of their agronomic characteristics, to lead to soil erosion and coincidental runoff or leaching of fertilizers and pesticides into water systems.

The policy influence is often surprisingly specific (see Table 1.2.). Under the American system of deficiency payments, for instance, farmers have little incentive to rotate subsidized crops with grass, alfalfa or other soil-conserving uses; rather they are strongly encouraged to maintain their 'base acreage' of cropland on which their eligibility for future payments is calculated. Given that over two-thirds of all cropland was enrolled in commodity programmes by the late 1980s, this bureaucratic requirement had a decisive land use effect, preserving the area of cropland and preventing farmers putting land to fallow or into a non-subsidized break crop. Creason and Runge (1990) quote a study by Young and Painter (1990) which examined the impact of federal commodity programmes on farmers' planting decisions, underlining the large opportunity costs of diversifying crop mixes and planting green manure rotations when deficiency payments are high and hectarage reduction programmes

Table 1.2. Independent short-run effects of agricultural policy on environmental quality. Source: Reichelderfer (1990).

Agricultural policy instruments that:	Net effect on:			
	Total soil erosion	Loss of wildlife habitat	Rates of agrochemical use	Total use of agrochemicals
Raise commodity prices	↑	↑	↑	↑
Tie farm income support to production levels	↑	↑	↑	↑
Reduce risk	↑	↑	↓	↑
Subsidize credit	↑↓	↑	↑	↑
Require short-term land retirement	↓	No effect	↑	↓
Establish cosmetic standards	No effect	No effect	↑	↑

Note: arrows indicate direction of net effect (increase or decrease) and do not imply whether the effect is 'good' or 'bad'.

are in force. In fact, the set-aside requirement intensifies the pressure on farmers to maximize the yields from land which remains in production, studies such as those by Hertel (1990) showing that the choice of hectarage controls is the choice of a high yield/high input system of agricultural production. Studies consistently show that the concentration of larger amounts of chemicals on fewer planted hectares as a result of set aside exacerbates environmental pollution problems. Just *et al.* (1991, p. 276) simulated the impact of deficiency payments and set aside for wheat and corn irrigation and groundwater depletion to conclude that 'agricultural commodity policy has contributed to structural change and input adjustments in the sector by stimulating output of programme crops, tying programme benefits to production and encouraging adoption of technical improvements ... Even though input substitution would have occurred in the absence of commodity programmes, the existence of such programmes has encouraged the substitution of chemicals for land as well as a more intensive use of that land'.

The consequence of this evolving complex of agricultural technologies and policies is a soil erosion problem perceived to be so severe that it would head the list of USDA's resource concerns as late as 1989 (USDA, 1989), despite a 50-year government programme designed to bring it under control (see Chapter 3). In the Great Plains particularly, wind erosion was still regarded as an issue of national concern by the Office of Technology Assessment (OTA) when it commissioned an expert panel to carry out a review of agri-environmental problems in 1995, drawing on the best available scientific knowledge (OTA, 1995b). Elsewhere, the planting of crops on steep slopes and in medium-textured soils continues to give rise to wind and water erosion throughout the Mid West and in the Southern Piedmont. Over 48.5 Mha of US cropland is estimated to be vulnerable to soil losses that exceed the rate of soil creation (OTA, 1995b). A traditional justification for concern has been the implications of soil loss for productivity and food supply (see, for instance, Larson *et al.*, 1983) but from the early 1980s onwards it became clear that the so-called 'off-site' damage of erosion – what happens to the soil when it leaves the field and enters streams, rivers and lakes – may be much more significant in environmental terms. An important study conducted by Pierre Crosson and a co-worker at Resources for the Future, a Washington-based think-tank, in the early 1980s, had already begun to question the conventional wisdom that agricultural sustainability was actually threatened by the erosion rates then being experienced, concluding that technological advance had largely compensated for any productivity reduction due to soil 'loss' since the 1940s (Crosson and Stout, 1983). According to Crosson's estimates, even if US cropland continues to erode at 1983 rates for another 100 years, yields will be no more than 3–10% lower than without any erosion. As he later put it 'these losses are so small as to be virtually lost amongst the statistical chaff' (Crosson, 1996, p. 4). On the other hand, while the USLE and the Wind Erosion Equation used in the 1977 and 1982 NRIs estimated the physical loss of soil on farms they said nothing about where the soil had gone. In 1985 the Conservation Foundation calculated that the damage done to commercial fisheries, recreation and other interests from the sediment which finds its way into watercourses was worth between $3 and $13 billion in 1980 (Clark *et al.*, 1985), of which a third was attributable to agriculture. Ribaudo (1989) later re-estimated the aggregate off-site costs of erosion from all sources at $8.8 billion annually (1986 prices). Taking the $4.5 billion of this total thought to be due to agriculture, Crosson (1996) calculates that this is equivalent to 25% of net farm income in 1986, net of government payments. At about this time, USDA research was also beginning to point to the large costs of eliminating chemical contamination due to pesticides from rural wells supplying domestic water (OTA, 1995b). Evidence that agricultural pesticides and nitrates were reaching aquifers began to emerge in the 1970s, when surveys by the Environmental Protection Agency (EPA) detected levels of nitrate in up to 60% of the wells sampled (US Environmental Protection Agency, 1990b). Although there have been no comparative surveys of groundwater quality in the US, researchers have discovered significantly higher pesticide residues and concentrations of nitrate in the most intensively farmed parts of the country. The US Geological Survey found that nitrate concentrations in groundwater along the South Platte river in Colorado, for instance, had exceeded public health standards for over 20 years (OTA, 1995b). In the surface waters of the Great Lakes, concentrations of toxic chemicals were still high in the early 1990s, despite considerable success in eliminating contamination from industrial sources (OTA, 1995b). Several large-scale studies show that agriculture is a significant source of the nitrogen, phosphorus and sediment found in the nation's surface waters (see Table 1.3). Crutchfield *et al.* (1993) estimating that 50% of the nutrients currently reaching freshwater systems is due to agricultural runoff. The problem remains especially acute in the Corn Belt, where fertilizer and pesticide residues are widespread and EPA water quality standards are regularly breached (OTA, 1995b). Indeed, pollutants from Corn Belt agriculture have been tracked all the way down the Mississippi, ending up in Louisiana's Gulf Coast estuaries (OTA, 1995a) where they are thought to have a range of damaging environmental effects. The Chesapeake Bay estuary on the east coast continues to suffer from eutrophication due overwhelmingly to pollution emanating from agricultural sources (Kahn and Kemp, 1985) and fertilizer used in the sugar fields of south-central Florida and pollutants from dairy farms have been implicated in the eutrophication of the Everglades.

Compared with their European counterparts, US agri-environmentalists have never been particularly exercised by the implications of agricultural intensification for the farmed landscape and wildlife on farms. As Westmacott (1983) observes, the most visible agricultural change during the 1970s and 1980s in rural America was the replacement of grassland prairie with cropland and while the result was an undoubted simplification of the landscape

Table 1.3. EPA assessment of US surface water quality, 1992. Source: OTA (1995a).

Water	Total resource base[a]	% assessed	% impaired	% fully supporting designated uses	Rank of agriculture as source of pollutants
Rivers and streams	5.63 million kilometres	18	38	56	1st – primary source
Lakes, ponds, reservoirs	16.2 million hectares	46	44	43	1st – primary source
Great Lakes shoreline	8660 km	99	97	2	NA[b]
Ocean shoreline	90,300 km	6	14	80	NA
Estuaries[c]	95,500 km²	74	32	56	3rd – notable source[d]
Wetlands	112 million hectares	4	50	50	1st – primary source

NA, not available.
[a] Contiguous US and Alaska.
[b] Atmospheric deposition is ranked first.
[c] Not including Alaska.
[d] Municipal point sources and urban runoff are ranked first and second.

in biological terms, there was no particular sense of a farmed landscape in retreat. Moreover, the nature of agricultural change was arguably less destructive of landscape features because:

> most farmers have found it possible and more cost effective to purchase additional land than to invest in costly reclamation schemes (as in Western Europe). Therefore natural features and obstructions tend to remain intact, though in some landscapes these are anyway few and far between. Prairie landscapes have not changed dramatically as a result.
>
> (Westmacott, 1983, p. 11)

Nevertheless, by the 1980s a debate had been joined about the biodiversity impact of wetland conversion and the ploughing of native grasslands, the issue coming to be seen to be as important as soil erosion by the time of the 1995 OTA assessment. The US Fish and Wildlife Service conducted a survey of wetland trends between the mid-1950s and the mid-1970s which showed that the total wetland area had declined from 43.7 Mha in 1954 to 40 million in 1975 (Tiner, 1984). According to estimates made by Heimlich and Langner (1986), 87% of this loss was due to agricultural conversion. Furthermore, a US Department of the Interior (1988) report concluded that various federal programmes were fuelling this process and had been since the 1930s, particularly in the Mississippi Delta and Prairie Pothole regions. In areas like the Mississippi Valley the surviving wetland area and 'bottomland forest' have been further degraded by the effects of agricultural pollution reported above. With the formation of the National Wetlands Forum in 1987 (Conservation Foundation, 1988), came recognition of the environmental importance of the wetlands resource and a campaign to prosecute a 'no net loss' strategy. Most of the native short-grass prairie once found in the Prairie Pothole area of Minnesota and the Great Plains had been converted to farmland by the late 1980s, the areas remaining being increasingly fragmented. The 'plough out' of the mid-1970s, which resulted in the conversion of up to 99% of native grassland in some locations, is now recognized as one of the largest single human-induced reductions in a North American ecosystem ever experienced, with consequent steep declines in grassland birds and other species of the open plains such as prairie dogs and the swift fox (Harrington, 1991). The less dramatic, but even more widespread, decline in the biodiversity of intensively farmed land is a more recent concern in the US. As the OTA (1995b, p. 20) remarks 'the net effect of changes in land use [here] – fewer kinds of wildlife species and a decreasing ability of the landscape to support common wildlife species – may appear insignificant at farm level, but are striking in a large, intensively cultivated region'. Protecting wildlife on farms, it would seem, had emerged as a new priority for US policymakers.

Agricultural Change in the European Garden

Agri-environmental change in Western Europe has followed a broadly similar path, but from a point very much further back in technological terms. As discussed above, with one or two notable exceptions such as The Netherlands' intensive livestock sector and cereal production in the Paris basin, continental agriculture after World War II was under-capitalized and still overwhelmingly peasant based, with any technological improvement in the inter-war period limited to a minority of farms. Over the next three decades, however, European agriculture would undergo a remarkable transformation, in Coppock's (1963, p. 45) words 'changing from a kind of handicraft to an industrial operation'. To be precise, the modernization process occurred in two stages (Gardner, 1996): an early restructuring stage, during which farm size increased, many marginal peasant farms disappeared and modern methods of livestock and animal husbandry were widely adopted; and a later stage, when even more advanced techniques were taken up by the most productive farmers. Again, it is difficult to disentangle the role of government policy and the CAP from the effects of endogenous technical progress in all this. As Cheshire (1985) points out, though, the fact that the invention of the CAP in the early 1960s created an economic environment offering price levels undreamt of by most farmers cannot be immaterial, the most convincing explanation of recent environmental change on farms boiling down to profit maximization in the face of factor substitution and (partially policy-induced) changing relative prices. To start with, economic theory suggests a positive relationship between product prices and output which chiefly operates in the short run by encouraging a greater use of variable inputs (Potter, 1995). As in the US, it would seem that farmers' immediate reaction to a coupled system of price support was to intensify production by applying more non-land inputs to every hectare of land in production. Between 1970 and 1990 applications of nitrogen fertilizer in the original EC-6 increased significantly, with later members like Ireland and Greece experiencing dramatic increases in use (see Table 1.4). Stanners and Bordeau (1995) estimate that utilization rates overall increased by over 75% between 1970 and 1989. The large-scale use of pesticides is similarly a phenomenon of the last 30 years and, from a low base, rates of increase have been dramatic. In France, for example, total annual pesticide use increased from 25,000 to 100,000 tonnes between 1971 and 1981 (Agra-Europe, 1991). The revolution in farming techniques which accompanied this in-tensification meant an increase in the scale of production and a displacement of labour by machines. A 30% reduction in agricultural manpower between 1950 and 1990 was achieved through a steady rise in the number of tractors and other machines in use on Europe's farms and a hefty increase in energy consumption (CEC, 1995). To the extent that this substitution was a goal of policy, encouraged directly through subsidies designed to lower the real user cost of capital, the CAP is heavily implicated here, and in the trend towards specialization which now began to take root.

Beginning in the 1960s, farmers in north-western Europe started to abandon the mixed farming systems which had traditionally been adopted to

Table 1.4. Average nitrogen fertilizer application rates in the EU, 1970/1990[a].
Source: Burch *et al.* (1997).

	Average nitrogen usage (kg ha^{-1})		
Country	1970	1990	% change 1970–1990
Belgium & Luxembourg	106	125	+18
Denmark	99	142	+43
France	53	94	+76
Germany (former West)	88	129	+47
Greece	24	85	+248
Ireland	18	92	+408
Italy	36	63	+87
The Netherlands	188	196	+4
Portugal	N/A	58	–
Spain	N/A	48	–
United Kingdom	66	122	+62

[a] Figures show average application rates for arable and improved grassland.

spread risk and maximize the 'joint economies' of having crops and livestock on the same farm. This was because economies of scale in the use of the specialized machinery and equipment that were rapidly becoming indispensable in farming operations could only be achieved with high rates of throughput. Under the CAP, however, opportunities for amalgamation to increase the overall scale of operations on individual farms were relatively few, for despite a steady diminution in the number of holdings, the high price regime had the effect of keeping more farmers on the land than would otherwise be the case (Colman and Traill, 1984). The result was that farmers reduced the number of enterprises on their farms in an effort to realize what Bowers and Cheshire (1983) call 'economies of specialization', 'the induced investment and employment loss (under the CAP) leading logically to the choice of a specialized enterprise' (Cheshire, 1985, p. 15). An indicator of this is the decline in the average number of enterprises on farms: between 1968 and 1974 these declined from 3.18 to 2.85 in the UK (Britton, 1977). The favoured enterprise was cereals, which expanded in area throughout the 1970s and 1980s. Bowler (1985) reports that cereals have become so central to farming in Western Europe that they are now grown on over 60% of holdings, with large monocultural concentrations in central and northern France, eastern Britain, Denmark and parts of southern Italy. Throughout northern Europe, but especially in the UK, Denmark, Belgium and northern France, the removal of hedgerows to facilitate large-scale and more specialized farming operations was the first visible sign of gathering agri-environmental change. Here, the displacement of mixed farming centred on a bocage-type landscape of small enclosures and the accompanying outward expansion of intensive arable production, slowly began to attract public concern. In the UK, publication of the Countryside Commission's 'New Agricultural Landscapes' study in 1974 (CC, 1974), brought home the extent to which agricultural intensification was sweeping away many of the traditional features of the enclosure landscape. Later, the Nature Conservancy Council (NCC) made this bleak assessment of the prospects for wildlife in Britain's countryside under the CAP:

> While a few habitats that are rich in wildlife are increasing, most in the intensively farmed parts of Britain are declining in size, in quality, or both. The decline is serious: it is occurring throughout the lowlands and more fertile uplands of England, Wales and Scotland ... the rate and extent of change during the last 35 years have been greater than at any similar length of time in history.
> (NCC, 1977, p. 21)

During the late 1970s hedgerow loss in the UK was running at an estimated 28,000 kilometres per year, with similar rates being reported from sample surveys in other member states (Baldock, 1990; Potter *et al.*, 1996). In Brittany and Normandy, for example, the disappearance of hedges, copses and lines of trees was a common experience in these years. More widely, land consolidation schemes undertaken to streamline farming operations meant fewer hedges, trees, ditches and field boundaries and the creation of the flat, homogenized agricultural landscape now so characteristic of large swathes of northern continental Europe. As Turner (1980, p. 1) observes, the effects of *remembrement* duplicate the landscape changes that had already begun to sweep through much of lowland England, with the inevitable result that 'output soars but the environment suffers'.

By this time, the ecological effects of intensification were also becoming increasingly clear; UK ecologists were among the first to identify a link between the application of persistent organochlorines and declines in populations of raptors and other species (Sheail, 1985). In a direct sense, pesticide contamination, eutrophication of watercourses and spray drift have impoverished habitats and extinguished species on a broad front. An example of this is the steep decline in up to 75 species of arable weeds in European cornfields as a result of herbicide applications (Burch *et al.*, 1997). At the same time, the wider use of nitrogen fertilizers, by enabling less fertile land to be brought into production, has contributed to the loss of semi-natural vegetation and wildlife habitat on lowland farms. Baldock (1990) documents the steady decline in the extent of habitats such as wetlands, semi-natural grassland, woodland and heathland throughout the 1970s and 1980s in northern Europe as farmers searched for ways of boosting production in order to maximize their receipts of price support. In Belgium, for instance, over 1000 ha were being drained annually in these high change decades, while The Netherlands lost 55% of its wetland area. Losses of unimproved grassland have been just as severe. A Council of Europe study of grassland estimated that it had been reduced to 10% of its former area, chiefly due to agricultural improvement. For the UK, NCC (1977) reported losses or serious damage to 95% of lowland unimproved grassland, 80% of calcareous grassland and 40% of lowland heath between 1949 and 1984. The cumulative effect of all this, most commentators agree, had been a general impoverishment of the lowland farmed landscape. Habitat loss and fragmentation has meant that three species of flowering plants, up to four species of dragonfly and one butterfly species have become extinct in Britain since 1947 (Green, 1986). In Germany, agricultural intensification and specialization has been identified as the primary cause of declines in up to 400 species of vascular plants (Baldock, 1990). The 'Red List' of endangered species shows that 30% of flowering plants, 40% of bird species and 50% of mammal species are either extinct or endangered in Germany, principally because of agricultural change (Agra-Europe, 1991). As a recent British study concluded, the legacy of decades of intensification and landscape change is a situation in which any wildlife interest which remains in intensively farmed countryside has been pushed back into small refuges 'to exist as fragments of habitat in the cracks between commercial land uses' (Adams *et al.*, 1992).

Nor has intensification been confined to arable regions. The equivalent process applied to livestock production is reflected in a steady climb in livestock numbers in northern member states over the same period as price support and, in the uplands, the combined inducements of the sheepmeat regime and the Less Favoured Areas Directive, encouraged farmers to reseed and reclaim grassland in order to support larger flocks and herds. In Ireland, sheep numbers tripled in just 8 years during the early 1980s, while in the UK they rose by 79% between 1985 and 1992 (Beaufoy *et al.*, 1994). Areas of 'rough grazing' have correspondingly declined, up to 150,000 ha being reclaimed in the UK between 1978 and 1993, for example (Barr *et al.*, 1993). In Ireland, 'improved' permanent grassland has been advancing at the expense of rough grazing at the rate of 4000 ha a year since the mid-1980s. Paradoxically, nature has suffered as much from the abandonment of land and declines in management as from intensification in many marginal upland areas. A more intensive stocking of the better land has typically been accompanied by a trend towards the ranching of the more remote parts of a farm and a decline in traditional land management practices such as burning and grazing of heather moorland. There has also been a shift away from over-wintering stock on the hill in favour of keeping them in the in-bye. As Felton and Marsden's (1990) study of heather moorland in the UK illustrates, this has led to habitat decline and biodiversity loss often as significant as that due to intensification and over-utilization. In France, it is estimated by Delorme (1987) that over 3 Mha of agricultural land have been abandoned since 1970, many of them in upland and mountainous regions because of changes in grazing patterns, though Baldock *et al.* (1993) caution that this has been a highly variable process geographically. What can be said is that the areas of vineyards, orchards and other permanent crops declined by over 40% in the four decades after the war, mostly due to the marginalization of traditional land uses (Potter, 1997). In many parts of central Europe and Scandinavia, the running down of farms and the withdrawal of management leads to the afforestation of the landscape as abandoned land succeeds to scrub and eventually high forest. Alternatively, land no longer farmed may be planted to a cash crop of trees. Vail *et al.* (1994) report that in Sweden the continuing replacement of traditional open agrarian landscapes with spruce forest has been one of the key stimuli to public debate about the environmental impact of agricultural change.

A broadly similar process of intensification on the best land, coupled with the marginalization or slow decline of traditional farming practices elsewhere, can now be seen taking place in the agricultural landscapes of the south (Potter, 1997). The accession of Spain and Portugal to the EC in 1986 gave a further boost to the already rapidly developing farm sectors in these countries, leading to a restructuring of production and the marginalization of what have since come to be regarded as 'high natural value' (HNV) farming systems (Beaufoy *et al.*, 1994). According to Extzarreta and Viladomiu (1989), it was in the open Mediterranean landscapes of central Spain particularly, that the increased irrigation of arable land triggered a more general intensification of production from the late 1970s onwards. Irrigation obviates the need for fallowing, with the result that less than 30% of arable land was left fallow in Spain in 1994 compared with well over 50% in the mid-1970s (Suárez *et al.*, 1996). This varies somewhat by region, with fallowing becoming most restricted in Castilla y Leon where it now extends to just 15% of the cultivated area. The agricultural area irrigated from the water-table doubled from 432,000 ha in 1962 to over 800,000 ha in 1983 (Suárez *et al.*, 1996). Dryland production in Spain and in Portugal is significantly more intensive than it once was, monocultural systems steadily displacing the traditional rich mosaic of arable crops, vines, olives and other permanent crops that offer such a rich habitat for steppeland birds. Baldock *et al.*'s (1993) evaluation is that, while there are still large tracts of countryside where low intensity arable production prevails, further intensification is inevitable in all but the most unproductive regions. It has been estimated that between 4 and 9 Mha were still being managed along traditional dryland production lines in 1992, but that under the National Hydrological Plan up to 600,000 ha of this will be irrigated and converted to intensive production over the next decade. In areas with very arid conditions and poorer soils, meanwhile, land abandonment is a real possibility. Other environmentally valuable permanent crops such as traditional olive groves which survive in Greece, Italy, Spain and Portugal are increasingly neglected or even abandoned. There are thought to be over 0.6 Mha of olive groves remaining in Greece, for example, many of them vulnerable to neglect under changing agricultural practices and patterns of land holding and occupation. According to Pain and Pienkowski (1997), olive production in Spain is already economically marginal and liable to be replaced on the better land by a more intensive crop. Portuguese montados and Spanish dehesa, intricate systems of farming which involve grazing within open woodland and between scattered trees (usually cork and holm oaks), together with small-scale, extensive arable production, continue to be major depositories of biodiversity in southern landscapes and the major remaining strongholds of biodiversity within the EU as a whole (see Table 1.5). In Spain there is an estimated 3.5 Mha of wood pasture but significant declines are now being reported in Extremadura and western Andalucia due to a complicated pattern of under-management and local increases in stocking densities (Díaz *et al.*, 1996).

The increase in livestock intensification is well advanced in southern member states, reflected in the same concentration of production on the best pasture at the expense of the abandonment or under-stocking of more marginal grassland elsewhere. In Extremadura, average stocking densities increased from 1.5 to almost three sheep per hectare between 1986 and 1991, causing widespread over-grazing and a decline in the quality of pasture for stock. The wider use of bought-in feed in recent years has brought about profound changes in the pattern of grazing, with stock that were once moved between winter and summer grazings now being kept on lowland pasture all the year round. Transhumance, the long-established seasonal movement of

Table 1.5. High natural value (HNV) farmland in Europe. Source: McCracken and Bignal (1995).

	Land areas in Mha			HNV farmland as % of country UAA	% share of total EU HNV farmland
	Land surface	UAA	Farmland under HNV systems		
France	54.7	31.0	7.7	25	13.7
Greece	13.2	9.2	5.6	61	9.9
Hungary	9.3	6.5	1.5	23	2.7
Ireland	7.0	5.7	2.0	35	3.6
Italy	30.1	22.6	7.1	31	12.6
Poland	31.3	19.1	2.7	14	4.8
Portugal	9.2	4.5	2.7	60	4.8
Spain	50.5	30.6	25.0	82	44.4
United Kingdom	24.4	18.4	2.0	11	3.6
Total	229.7	147.6	56.3	38	100.0

stock between summer pasture and winter grazings, is far less widespread than it once was throughout southern Europe (McCracken and Bignal, 1995). Its decline is having two categories of effect. First, grazing pressure is being increased on grassland in the plains and valley bottoms as stock are retained in the lowlands all year round. The grass steppes of central Spain, previously grazed at low densities and of great importance for birds including Montagu's harrier *Circus pygargus*, the little bustard *Tetrax tetrax*, and the globally threatened great bustard *Otis tarda*, are now beginning to exhibit vegetation changes brought about by over-grazing (Pain and Pienkowski, 1997). And in an illustration of the interconnectedness of agri-environmental change here, there is also evidence that the presence of greater numbers of livestock all the year round is disturbing the agro-ecology of agro-silvo-pastoral systems such as the dehesas mentioned above. Second, the decline of transhumance is threatening the mountain pastures themselves, this time due to under-grazing. It is estimated that up to 1.5 Mha of mountain and alpine grazings are used by mixed flocks of sheep and goats in the summer months in the Italian Apennines, for example. As Pain (1994) argues, their increasing abandonment should be a matter of great concern given their continental importance as reservoirs of biodiversity.

Compared with these rather complex problems of countryside management and under-management, 'New World' concerns about soil erosion and the public health effects of agricultural pollution have been slower to appear in European, and especially UK, public debate. Nevertheless, by the late 1970s the pollution impact of intensive livestock production in The Netherlands was beginning to attract government attention. With a huge 'manure surplus' being generated on specialized livestock farms, the problem was primarily one of safe storage and disposal. By the end of the decade it was estimated that only 45% of the 19 million tonnes of animal waste produced annually could be disposed of on the farms concerned, with a consequent danger of eutrophication and contamination of surface drinking water (Dubgaard, 1993). In Denmark it has been estimated that an investment of over 350 million ECU was required in the early 1990s to provide proper storage and treatment facilities for animal manure (Tamminga and Wijnands, 1991). In the UK and other parts of the EU, the growth in dairy herd sizes and the trend towards housing stock in cubicles rather than bedding them on straw, greatly increased the volume of liquid slurry to be disposed of, and with it the risk of serious pollution of surface water (Lowe *et al.*, 1990). Elsewhere, arable farming quickly emerged as the greatest non-public source of water pollution in the EU, Dubgaard (1993) calculating that agriculture's share of nitrogen emissions into surface water due to runoff and leaching ranged from 37 to 82% in most EU member states. Nitrate pollution of groundwater has been a particularly important issue for Europeans, particularly in Germany and France where Conrad (1990b) points to its rapid movement up the political agenda in the mid-1980s once the public health implications became clear (see Chapter 5). It has been estimated that 25% of EU groundwater supplies exceeded the World Health Organization (WHO) recommended limit in 1990 (Conrad, 1990b). Finally, soil erosion in the EU is primarily a problem of Mediterranean regions, where a combination of steep slopes, fragile soils and dry climatic conditions have produced rates of soil loss in excess of the 10 t ha^{-1} regarded as tolerable (von Meyer, 1988). Over 44% of the surface area of Spain is affected by soil erosion of some sort compared with 20% of Greece, 10% of Italy and 1% of France (Burch *et al.*, 1997).

While the precise nature of the interaction between technical progress and agricultural policy may still not be fully understood, this survey suggests that there is now sufficient evidence to suggest a strong policy-driven (even if not policy-induced) explanation for agri-environmental change in industrial countries since 1945. Inevitably, perhaps, it was never very likely that there would be a single convincing explanation of post-war environmental change. That there has been a powerful transformation of the relationship between agriculture and the environment is undeniable. There would also be broad agreement that, while farm policy may not have instigated all of the changes in farming practice and land use, it has powerfully reinforced the trend towards greater intensification and specialization. It would appear that the 'coupled' system of farm support that is a common feature of both the US and the EU has provided an irresistible economic incentive to expand output and to do so in a land saving way. Harnessed to a technological revolution in farming, the result is the rapid intensification, specialization and concentration of production which lies behind many of the agri-environmental problems reported above. In the US, the peculiarly American combination (at least until recently) of commodity price support with compulsory land retirement has had the effect of simultaneously braking and accelerating agri-environmental change over a period of 40 years. Soil erosion re-emerged as an issue of public concern following the plough-out of the middle 1970s, but its persistence when land was later being removed from production suggests that longer term changes in the pattern and intensity of production, many of them still linked to policy, are ultimately to blame. To soil erosion has since been added the problems of water pollution and wildlife decline, public debate shifting its focus away from the New Deal notion of soil erosion as a threat to the productivity of the land in favour of a more comprehensive definition of the environmental impact of modern farming. This redefinition of the problem, largely the result of an improving knowledge base and a growing appreciation of the externality costs of production, had important implications for the way agri-environmental reform would later be justified and conceived here.

The impact of the EU's CAP on the environment has been just as complex. The decision to follow a farm survival policy which attempts to solve the income problems of large numbers of marginal producers by encouraging more production on the back of modernization rather than through restructuring, has made a deep impression on the ecology, appearance and character of the European countryside. It is debatable whether the streamlined industry of fewer, larger farms that Mansholt had in mind would have been any more environmentally friendly than the polarized one which currently exists. However, there is something in the argument that the CAP has brought about changes to the agricultural geography of the EU and to the pattern of production which have been especially damaging environmentally in a European context. The decline of mixed farming and the advance of intensive arable production throughout north-western Europe in the 1970s, for example, gave rise to many of the changes in the agricultural landscape that first stimulated public concern. Agricultural pollution, landscape change and the loss of bio-diversity due to the intensification and concentration of production in some locations offset by marginalization and even land abandonment elsewhere, remain very European concerns. The recent intensification of dryland arable production in southern member states, combined with the decline of traditional, high natural value systems of farming, is merely the latest manifestation of this paradox of agricultural change.

2 The Pressures for Reform

If policy change requires a consensus about the problem at hand and its causes, then by the early 1980s conditions were ripe for doing something about the environmental effects of agricultural support. Agri-environmentalists in both the US and, within Europe, the UK, could point to an expanding body of research documenting the erosion of soil, pollution of water, loss of biodiversity and degradation of landscapes which had invariably accompanied agricultural expansion since the war. In the US, the RCA of 1977 had directed the USDA to appraise the status of the nation's soil, water and related resources in order to prepare a programme for furthering their conservation. Even as the RCA process was getting under way, a number of other studies were coming on stream. The National Agricultural Lands Study had investigated the conversion of farmland to non-farm uses and made calculations about the likely impact of this process (large-scale and long established) on agricultural output and productivity, while the NRI had already been completed by the Soil Conservation Service (SCS) which documented the use and quality of the farmland resource which remained. This latter was now providing the first authoritative estimates of the scale of the soil erosion problem, furnishing lobbyists with the data they needed to press for an overhaul of federal soil conservation programmes. It was an information revolution which transformed the conservation debate and set in train new thinking about how to rebalance the national soil conservation effort. The effect was not only to put soil erosion into some sort of national perspective but also to improve understanding of the nature of the problem and, critically as it turned out, its geographical distribution. British conservationists had, by comparison, a far less complete picture of environmental change, relying on anecdotal evidence and sample surveys like those commissioned by the Department of the Environment (Huntings Surveys, 1986). Early work, such as the 'New Agricultural Landscapes' study mentioned above, used case studies to identify the types of change that were taking place, saying little about the extent or rate of change. In a review of the literature conducted in the early 1980s, Munton (1983) could still point to significant gaps in knowledge which made it difficult to assess just how rapidly change was taking place at a national level. Comparison between locations was frustrated by the lack of any standardized methodology and many landscape types and situations had not been surveyed at all. It was not until 1984 that the Nature Conservancy Council began to put together a systematic record of habitat loss and species decline – though the published projections, when they arrived, were dramatic, showing that since 1949 80% of limestone grasslands, 50% of lowland mires, up to 50% of ancient woodland, 40% of heathland and 30% of upland moorland had gone (NCC, 1984). By this time the Institute of Terrestrial Ecology (ITE) was using its land classification system as a sampling frame to carry out more detailed land cover and ecological surveys throughout Britain. The Countryside Survey 1990 provided the first comprehensive survey of environmental stock and change and laid the foundation for a repeat survey in 2000 (Barr *et al.*, 1993). Meanwhile, agri-environmental reform was well served by brilliant polemics like Marion Shoard's *The Theft of the Countryside* which dramatized the problem and brought home the rather incendiary idea that landscape destruction was being financed with public funds (Shoard, 1980). This struck a chord with a countryside-loving public that was by now being confronted daily with the obvious visual signs of what Green (1985, p. 231) regarded as:

> as sweeping as any in its history. It is easy to argue that change has always been accepted in the past and will be again, but the changes today are unlike any in the past in that they are producing a countryside barren of wildlife and destitute in its landscapes just at a time when people want and need a more varied rural environment.

Commentators and lobbyists in both countries only now began to make an analytical connection between the operation of government policy and these, their environmental effects. The construction of this common understanding was a more gradual, negotiated process but a critical one none the less for the agri-environmental reforms that were to come.

The Erosion of Traditional Thinking

It began in the US, where federal commodity support, far from being regarded as one of the causes of environmental degradation, was traditionally seen as a means of ensuring that more conservation got done. The conventional wisdom, firmly anchored in New Deal thinking, was that agricultural prosperity, underwritten by price support, was the best way to ensure farmers had sufficient capital to invest in the conservation of the soil. As Swanson (1993) points out, a basic assumption behind the New Deal conservation programmes enacted in the Depression was that erosion occurred, not because farmers were unwilling to invest in soil conserving practices, but because they could not afford to do so and that, provided the income crisis could be eased, farmers should return to environmentally sound practices. It was also asserted that the principal cost of soil erosion was incurred 'on site' and borne by the farmer himself in the form of lost productivity. More conservation would be undertaken if income-secure farmers could be persuaded to adopt a more telescopic vision of farming and their long-term future within it. In 'micro' terms, policy was underpinned by a definite view of farmer decision making which emphasized the costs and benefits of soil conservation to the farmer and saw the solution in terms of providing the financial means to undertake soil conservation investment while simultaneously encouraging farmers to apply lower discount rates in their assessments of such investments. This was consistent, at a deeper level, with a progressive conservation image of farmers as 'manager(s) of nature, extracting a bounty to support the continued prosperity of the nation' (Batie, 1986, p. 4).

To start with, however, the Agricultural Adjustment Act of 1933 concentrated on paying farmers to take land out of production, making no mention of soil conservation, even as an ancillary goal of policy. Producers were offered rental payments by USDA to take land out of production, and the federal government was empowered to enter into marketing agreements with processors in order to control farm product prices. All these activities were to be financed by a tax on processors. In January 1936 this policy was ruled to be unconstitutional by the Supreme Court on the grounds that economically constrained farmers had little choice but to enrol land and that 'the asserted power of choice [available to farmers] is illusory'. USDA's response was to propose making soil conservation the primary aim of the adjustment programme in order to emphasize its voluntary nature. Land would still be rented and withdrawn from production but this would be for the dual purpose of promoting conservation and reducing surplus production (Baker *et al.*, 1963). Thus, while the first aim of the Soil Conservation and Domestic Allotment Act of 1936, which set up what came to be known as the Agricultural Conservation Program (ACP), was still to improve farm incomes by restricting output, this would be achieved by shifting land from 'soil depleting' to 'soil conserving' crops like grass, legumes and green manures. The fact that the former happened to be crops most in surplus was a happy coincidence which nevertheless strengthened the case for commodity support and soil conservation being complementary policy goals. Payments were also to be made for installing conservation practices such as terraces on contracted land. Administrator of the Agricultural Adjustment Administration (AAA), Howard Tolley, assured the public that a 'wise use of the land is the only direct objective. Of course, shifting land from crops that deplete the soil to those that conserve and build soil will have a tendency to prevent big surpluses from piling up and reducing farm income', but this was incidental to the main policy thrust (quoted in Kramer and Batie, 1985, p. 310). Reichelderfer (1990, p. 210) comments that:

> as its title suggests, the Soil Conservation and Domestic Allotment Act of 1936 closely correlated soil conservation objectives with production adjustment goals. One could say, in fact, that it exploited soil conservation goals to achieve production adjustment goals ... Nevertheless, the programs established in 1936 introduced what became a long-standing practice of using resource conservation programs to complement agricultural market interventions.

With demand for food rising sharply during and after the war, and evidence emerging that myopic farmers would only undertake conservation if they were compensated generously to do so (see Heady and Allen, 1951), the emphasis of the ACP shifted away from land diversion into subsidizing the adoption of conservation practices on farms. Cost sharing of conservation practices like terracing, contouring, inter-cropping and laying down grass waterways now took place on a vast scale under the auspices of the USDA's Agricultural Stabilization and Conservation Service (ASCS), which ran the programme, together with research, extension and advice from the SCS. Between 1936 and 1979 expenditure on ACP increased from $60 million to $233 million, actually a reduction when adjusted for inflation but still impressive, particularly since the ACP in 1936 was discharging virtually all of USDA's production control responsibilities. SCS appropriations have expanded from $791,000 in 1936 to $523 million in 1980. By this latter date some 2.4 million farmers were in Soil Conservation Districts, voluntary associations of farmers with responsibility for overseeing the local implementation of federal policy (Rasmussen, 1982). As Rausser (1990) points out, the formation of these districts was a critical institutional investment in itself, allowing local farmers significant powers of discretion but also strengthening the legitimacy of federal soil conservation programmes. Government agencies like the SCS gained considerable credibility with farmers by showing themselves to be responsive to conservation district priorities (later this same closeness would be seen by environmentalists as part of the problem of under-performing soil conservation programmes). Policymakers temporarily resuscitated the idea of a conservation reserve in the mid-1950s when surpluses once more threatened to depress prices and under the Soil Bank programme of 1956 they offered farmers 3–10-year contracts to divert crops into conservation uses. Under the conservation reserve, participants received two types of payment: annual rental payments and cost-sharing payments for conservation investments. In Wayne Rasmussen's (1982) opinion this attempt to revive the conservation approach to agricultural adjustment and price support was moderately successful, diverting over 8.5 Mha into grass or trees under long-term agreements. 'It ended rather promptly, however, with a return to more traditional price support programmes and increasing emphasis on resource conservation, including water as well as soil, in SCS and ACP' (Rasmussen, 1982, p. 11). The Great Plains

Conservation Program, a more permanent scheme, offered farmers in drought-prone areas similar long-term contracts based on income and capital payments to divert cropland into grass. By 1969 some 34,000 landowners had enrolled more than 25 Mha, a further 16,000 having been attracted in by 1980 (Batie, 1984).

The consensus among conservationists that these programmes were a good thing was still more or less intact by the mid-1970s, despite rumours that not all the money was being used for strictly conservation ends. Thereafter criticism mounted steadily as the NRIs of 1977 and 1982 made clear that soil erosion was still substantial despite a heavy investment of public funds over many years. For the first time, there was talk of 'pork barrel' soil conservation programmes, functioning to serve the needs of farmers rather than those of the environment and maintained in existence by an 'iron triangle' of agency personnel, interest groups and members of the powerful congressional sub-committees on agriculture (Browne, 1988). Increasingly, legislators began to substitute their own judgements for those of the USDA and its agencies and there were acrimonious debates in Congress about the apparent failure of ASCS to fully prosecute its conservation mission (Rasmussen and Baker, 1972). At the same time urban conservation groups began agitating for action on water quality, an amendment to the 1972 Federal Water Pollution Control Act requiring states to identify where significant non-point pollution was taking place and to enact regulations to control it. Responding to the same sense of public disquiet, the Iowa Soil Conservation District Law was enacted in 1971 to allow the state authorities to declare soil erosion a nuisance which landowners had a legal duty to correct. Thigpen (1994, p. 75) comments that:

> the traditional conservation organizations and agencies found themselves in a dilemma. While their *raison d'être* was sound management of the nation's resources, they were in a position of opposing many proposed environmental laws because of disagreement with the means of accomplishing such management – in a word, regulation.

In 1976 Congress produced an oversight directive which required USDA to present evidence that conservation programmes were having an effect on soil erosion. The RCA of 1977 was the immediate result. This required USDA to undertake a systematic appraisal of the magnitude of the soil erosion and water quality problems and to evaluate the performance of conservation pro-grammes in the light of this knowledge. The effect was to raise the visibility of soil conservation as a national environmental concern and, critically, to require appraisals of programme performance as part of the budgeting process. Helms (1990, p. 14) comments that:

> Government observers [initially] scoffed at the prospect of another study as a way of evading a difficult issue. In retrospect, the RCA seems to have become one of the instrumental factors in passage of the conservation provisions of the 1985 Farm Bill. Previous studies of conservation needs by SCS had concentrated

on identifying conservation problem areas and needed conservation work. The studies started under RCA concentrated on quantifying soil erosion.

When, in 1977, a report to Congress by the Comptroller General entitled *To Protect Tomorrow's Food Supply – Soil Conservation Needs Priority Attention* (GAO, 1977) drew on these data to take ASCS to task for its poor targeting of conservation funds, analysts began to look again at the operation of these policy monoliths and the assumptions they enshrined. According to the Comptroller General, the programmes were failing for two reasons: (i) conservation sub-sidies were being given to farmers regardless of the severity of their soil erosion problem or potential for soil erosion; (ii) too many of the practices installed (such as drainage, land levelling and liming of fields) boosted production rather than conservation. Indeed the proportion of ACP-subsidized practices that were actually soil conserving had dropped from 59% in 1970 to only 45% in 1977, the rest merely enhancing output or improving farm productivity. As for conservation plans, 'many in SCS files were outdated, forgotten by the farmer or just not carried out or used as a basis for making farm decisions' (GAO, 1977, p. 7). A charitable interpretation of ASCS's scatter-gun strategy was that the principle that only profitable farmers undertook conservation had been taken too far; the more realistic interpretation was that money had been spread like jam to secure the support of as many farmers as possible. As one rather shrill critic, reflecting on the RCA estimate that up to 4.7 billion tonnes of soil may be eroding annually, put it:

> USDA platitudes, conservation rhetoric, mere lip service in the face of such appalling losses are even more shocking than the statistics. We have here a situation of a patient suffering from pneumonia, and the doctor responsible diagnosing the sickness as a common cold.
>
> (Barlowe, 1979, quoted in Batie, 1982, p. 28)

Later evidence from a 1980 assessment of the ACP indicated that considerable cost savings were achievable from a better targeting of the policy. As Table 2.1 shows, the distribution of cost sharing and technical assistance between states was failing to reflect the concentrated nature of the soil erosion problem. Even the American Farm Bureau was willing to concede that less than 5% of the total SCS budget was being used to finance soil conservation on land most vulnerable to erosion (Miranowski and Reichelderfer, 1985). Pressure grew to tighten up the targeting of conservation subsidies and Congress demanded evidence that dollars invested in the ACP yielded a commensurate improve-ment in soil conservation.

More significant still, there was growing awareness of the poor coordination between commodity and conservation programmes and of the powerful in-centives offered to farmers to put highly erodible land under highly erosive crops. The American Farmland Trust (AFT, 1984) commented that short-term land retirement programmes had done little to reduce soil erosion, despite the popular conceit that 'they killed the two birds of overproduction and soil

Table 2.1. Distribution of 1983 US soil conservation expenditures relative to the soil erosion threat. Source: based on data from USDA (1985).

Region	% of federal soil conservation expenditures	% of total cropland with erosion exceeding 4 t ha^{-1} year^{-1}	% contribution to US erosion
North-east	8.7	2.8	2.3
Lake States	8.0	8.5	8.3
Corn Belt	18.4	26.0	25.4
Northern Plains	10.8	14.5	17.0
Appalachian	13.1	6.0	6.0
South-east	10.6	3.2	1.7
Delta States	7.0	2.9	4.1
Southern Plains	12.3	17.1	19.4
Mountain States	5.6	14.2	12.6
Pacific	4.3	3.8	3.3

erosion with one stone', and in a report published in the same year, the Conservation Foundation pointed to the competitive nature of commodity support and soil conservation programmes, arguing for better targeting and co-ordination of the national soil conservation effort (Batie, 1984). Indeed there was evidence that the operation of the 'base acreage' system had brought many hectares of vulnerable farmland into continuous cropping, creating subtle but powerful incentives in favour of intensive production. Experience with PIK confirmed this. When it was first announced, USDA went out of its way to emphasize the soil conservation benefits of taking so much land out of production, the ASCS Chief predicted that up to 235 million extra tonnes of soil could be retained on farms because of the PIK programme (Cook, 1983). In the event, soil savings were small, farmers failing to install effective conservation covers on the 33.2 Mha of land that were retired and in many cases actually intensifying production on the land which remained in production. For commentators like Kenneth Cook (1983, p. 476), this typified the poor co-ordination of the conservation and commodity arms of federal government programmes, 'the failure to leverage more enduring conservation benefits with a $12 billion public investment [being due] to a singular lack of foresight, creativity and commitment at the highest levels of USDA commodity and conservation policymaking'. It also put a question mark against the ability of USDA to exploit the conservation potential of the larger agricultural policy changes now in prospect.

The reconversion of a large area to production in the following year confirmed these criticisms, and support now began to grow for a return to the conservation reserve of the Soil Bank experiment under which diverted land could be fixed in conserving uses for periods long enough to have a soil conservation effect (see for instance, Ogg and Zellner, 1984); there were even rumours that ASCS itself had begun to consider offering certain farmers multi-year contracts to set aside especially vulnerable land on a more permanent basis (Benbrook, 1979). But it was the idea of 'conservation compliance' which most rapidly gained ground in these preparatory years. The notion that farmers should be required to install conservation practices in order to qualify for commodity support or, under the 'green ticket' approach, in order to secure additional deficiency payments, first surfaced during public hearings conducted by SCS as part of the RCA process. Tom Barlow, a lobbyist for the Natural Resources Defense Council, took up the idea and during 1976 and 1977 was arguing for the imposition of a *quid pro quo* condition on all participants in the commodity programmes. It was probably the first example of sustained thinking towards the greening of US farm policy. In an important sequence of articles published in 1979 and 1980, Charles Benbrook proposed a 'Conservation Incentive Program' based on the wide but permissive application of conservation compliance (Benbrook, 1979, 1980). Under this arrangement, farmers would volunteer to have a conservation plan drawn up for their farm specifying crop rotations, land uses and conservation practices as appropriate. In return they would receive higher target and support prices for the duration of the contract. According to Benbrook, the advantage of this 'green ticket' scheme was that it got conservation on to a far larger number of farms than would ever be possible through the ACP, particularly given the latter's vulnerability as 'discretionary expenditure' to budgetary constraint (he assumed that conservation compliance would apply to 70% of cropland in the Mid West). As he put it 'The commodity programs are an appropriate delivery mechanism ... because of the potential to reach a high percentage of cultivated land and thereby support adoption of these practices' (Benbrook, 1979, p. 164). Whether this was the clinching argument so far as the National Association of Conservation Districts (NACD) was concerned is unclear; Kramer and Batie (1985) speculate that the rapid conversion of this key organization to the cause of conservation compliance during 1979 owed more to its anxiety to avoid regulation than a sudden conversion to agri-environmental reform. In any event, ASCS was now advocating conservation compliance and lobbying government for legislation to this end (Batie, 1986). Despite continued resistance to the idea from within ASCS ('cross compliance' had strongly negative connotations for farmers following its partial implementation under the 1977 Food and Agriculture Act), the Agriculture Secretary was instructed by President Carter to look carefully at the idea within the context of the RCA process.

> A major goal of your effort should be to analyze the implications of modifying or interrelating existing agricultural assistance programs to bring about greater reduction in soil erosion and related non-point pollution. The study shall also identify conflicts between farm income programs and soil conservation programs and develop recommendations for eliminating these conflicts where possible.
>
> (US Executive Office of the President, 1980, quoted in Kramer and Batie, 1985, p. 315)

When the RCA eventually reported late in 1980 it recommended that con-servation compliance be given careful consideration as a mechanism for extending the soil conservation effort on the nation's farms. The report acknow-ledged that a cross-compliance strategy would 'diminish the voluntary nature of existing USDA programs' but went on to argue that the case was still strong on both soil conservation grounds and in order to improve policy consistency. Serious debate about environmental reform of US farm policy had begun.

Defending Nature, Conserving Countryside

The debate in Western Europe at this point was nowhere near as focused or in-strumentally inclined. In countries like France indeed, there was barely any awareness that an environmental crisis in agriculture existed, let alone that agri-environmental reform was required. Buller (1992) comments that a strong agrarian tradition which equated farming expansion with the public good meant that environmentalists were hardly likely to challenge the agri-cultural industry on its environmental record unless presented with very convincing evidence to the contrary.

> Early French environmental concern made no distinction between agricultural occupancy and the rural environment. Centuries-old traditions of agricultural land management, many enshrined in the Napoleonic rural code, were regarded as essential to the protection of natural resource, but only to maintain the rural social and economic order. Hence, the emerging French concern for the rural environment was fundamentally different from that in Britain. The rural landscape was not separated, either aesthetically or functionally, from agriculture. Indeed, the notion of countryside 'preservation' as an end in itself, as found in the outlook of the Council for the Preservation of Rural England from the 1920s onwards, is contrary to the French belief in rural social and cultural sustainability.
>
> (Buller, 1992, p. 73)

As Chapter 5 will show, French agri-environmentalism did not therefore begin to surface until comparatively late in the day when research conducted by the French Ministry of the Environment in the mid-1980s into agricultural pollution and the contamination of water supplies from nitrates established that the problem was of sufficient concern to demand a government response. It would be later still that a connection would be made with the operation and procedures of the CAP. Nowicki's (1988) explanation is that environmental problems, where they were recognized, were much more likely to be related to the techniques of land reallocation and to the nature of decision making within state agencies than to price decisions made in Brussels. Not until 1987 would the CAP begin to make an appearance in the environmental press and then mainly in connection with the desertification risks of abandoning sup-port in Less Favoured Areas (Nowicki, 1988). In Germany, a major scientific report of 1985 by the Council of Environmental Advisors to the Federal Government (CEAFG) had catalogued the environmental impact of modern agriculture without offering any policy explanations or putting forward any CAP-related solutions (Höll and von Meyer, 1996). For the moment then the initiative lay with academics and lobbyists in the UK and, to a lesser degree, The Netherlands and Denmark, to create a climate of opinion conducive to agri-environmental reform. An important contribution to this end was made by British economists John Bowers and Paul Cheshire who, in a series of trenchantly written articles and books, argued the case for radical policy change on economic, financial and environmental grounds. Their central contention was that the economic inefficiency of the CAP is of a piece with its environmental effects and that 'the single most important change in agricultural policy from the viewpoint of conservation ... would be a reduction in the level of agricultural support' (Bowers and Cheshire, 1983, p. 138). Some time before, in 1969, they had argued, with great perspicacity as it turned out, that society could pay farmers to produce countryside instead of food, if it decided that this was a better use of public funds.

> Society does not owe farmers a living. There could be a case, however, for compensating farmers in so far as the community decides to hinder them from making a living. If, for example, society decides ... that hedgerows enhance the amenity value of the countryside by providing cover for wildlife and simple aesthetic charm, it might decide to offer farmers subsidies to grow hedges, not to grub them out.
>
> (Cheshire and Bowers, 1969, p. 15)

Now they deployed what Lowe *et al.* (1990) would later describe as the 'policy thesis' – the idea that agricultural policy is largely responsible for the post-war environmental degradation of the countryside because of the incentives it gives farmers to intensify production along capital-intensive lines. This was an important intellectual shift, moving the debate away from a focus on individual farmers and their separate acts of environmental destruction towards a more systematic analysis of the underlying driving forces (Potter, 1986). Indeed it was argued that the pace as well as the direction of post-war agricultural change was strongly policy determined, and in a very specific way, once policymakers had made the decision to increase the incomes of farmers through modernization rather than through structural change.

> The policy followed has involved the application of science to production within an existing distribution of farm size. This has involved specialization and intensification of production with the maximization of output per head and output per acre. This policy is the one that has maximized the damage to the environment and to the interests of other users in the countryside.
>
> (Bowers and Cheshire, 1983, p. 118)

Commentators like Davidson (1977), who argued that farmers simply needed to be better educated and advised, misconstrued the problem, it was claimed, by locating it in the minds of farmers. What was required was a radical re-structuring of farm support, not mere persuasion and appeals to the altruism of a few conservation-minded farmers.

Conservationists were by this time highly receptive to any proposal for agricultural policy reform, even of a type as potentially draconian as that being put forward by Bowers and Cheshire. An important coevolution in thinking was under way in which an awareness of the destructive potential of the CAP was matched by a growing recognition that the traditional conservation strategy of site safeguard and nature reserve and heritage landscape designa-tion was not working and could not be sustained. In 1977, the government's advisory agency on nature conservation, the NCC, had argued for the first time that agricultural expansion was seriously threatening the conservation resource in the wider countryside. It was an admission of the failure of post-war conservation to defend agricultural landscapes as a conservation resource and had profound implications for the future conduct of policy. Until this point, nature conservation had chiefly been seen as an enterprise of selecting and designating National Nature Reserves (NNRs) and Sites of Special Scientific Interest (SSSIs) 'to form a national network of areas representing in total those parts of Great Britain in which the features of nature, and especially those of greatest value to wildlife conservation, are most highly concentrated or of highest value' (NCC, 1984, p. 8). As Felton (1993) points out, the luxury of being able to present nature conservation as a largely scientific and educational project centred on the selection, management and protection of key sites, was only possible because a prescribed form of countryside management was not perceived to be required elsewhere. This was the message of the influential Scott Committee on Land Utilization in Rural Areas (Scott, 1942, p. 2), which had famously opined that

> farmers are unconsciously the nation's landscape gardeners ... even if there were no economic, social or strategic reason for the maintenance of agriculture, the cheapest way, indeed the only way, of preserving the countryside in anything like its traditional aspect would still be to farm it.

It was also a belief shared by prime movers for landscape conservation like John Dower, who argued that 'generally speaking, the interests of agriculture and of landscape beauty are at one' (Dower, 1945, p. 26). The Scott philosophy became the touchstone of post-war planning, justifying a dual approach to rural land use management which combined strict control over urban develop-ment under the town and country planning system to safeguard productive agricultural land, with an expansionist farm support policy bereft of any environmental safeguard. For the next 30 years, British conservationists would devote much of their resources and expertise to creating a conservation estate made up of national parks, NNRs, SSSIs and local nature reserves; they would be little interested, and have few powers, to influence the course of change on non-designated land. The prevailing image for nature conservationists was of a pyramidal countryside, with these key sites at its apex resting on a broad base of 'good wildlife habitat' assumed to exist elsewhere. Felton's (1993) comment is that this analogy, repeatedly used in official publications and promotional literature, merely reinforced the impression that agricultural landscapes, if not expendable, were of lesser value in conservation terms. They could safely be left to the farmers' care.

Now, with growing recognition of the destructive potential of modern agricultural practices, action was needed on a broader front to protect the conservation resource in 'the wider countryside'. Evidence from the NCC's surveys, showing that 15% of SSSIs visited in 1980 had suffered some damage or loss, in over 50% of cases because of agricultural intensification, suggested that site safeguard was failing, even on its own narrowly defensive terms. The conclusion had to be that 'however the results of this survey are interpreted, it is evident that damage to SSSIs is considerable and had reached a level that gives concern not only to the NCC but to all those who are interested in the future of Britain's natural heritage' (NCC, 1980, quoted in Adams, 1986, p. 99). Moreover, even where conservation sites could be protected from damaging agricultural operations like drainage, reclamation, reseeding and overstocking, they were still becoming increasingly isolated islands of habitat in seas of intensively farmed land. The insights of island biogeography and landscape ecology suggested that all but the most extensive key sites would lose their biodiversity and become ecologically vulnerable if they were not reconnected to the matrix of semi-natural habitats which existed when many of them were first designated. As Adams *et al.* (1992, p. 247) put it:

> The significance of SSSI loss is magnified because of the damage that has taken place in the wider countryside. Throughout lowland UK, sites of high wildlife value are limited in extent and fragmented. Around them is a much larger area of much reduced diversity and value. The survival of the wildlife interest of the SSSIs themselves depends to an extent on the nature of the landscape of the wider countryside. Thus the importance of SSSI loss reflects a wider policy failure, the solution to which lies beyond the policy of designating and protecting SSSIs.

The government's first instinct was to react to growing public concern throughout 1979 and 1980 by attempting to shore up the existing policy regime and strengthen site protection by requiring farmers to give NCC and other authorities advance notice of any intention to carry out damaging operations on SSSI land (Cox *et al.*, 1985b). Prior to this, a long-running dispute between farmers and conservationists in the Exmoor National Park in southwest England had underscored the need for enhanced powers of control over agricultural development, the national park authority finding itself unable to prevent the large-scale ploughing up of core moorland habitat, despite years of negotiation with local farmers (for a discussion of the Exmoor dispute, see MacEwen and MacEwen, 1982). Lord Porchester, appointed in 1977 by the Callaghan Labour Government to investigate the workability of voluntary management agreements between farmers and the national park authority, had recommended a system whereby farmers would have to notify the authorities of an intention to plough. Lump sum compensation would then be offered where applications were refused (Porchester, 1977). In anticipation of the later debate about the difference between compensating farmers for damaging operations withheld and subsidising the costs of conservation management, Porchester (1977, para. 10.20) comments:

> There is a fundamental difference between paying compensation to someone when he is denied the right to do something he wishes to do, and paying him to do something which he would not (necessarily) do of his own will or to his own benefit. We need to distinguish quite clearly between compensation on the one hand (which, in my opinion, should be paid on a once and for all basis) and payment for definable things required to be done beyond the farmers' own interest on the other (which could be paid annually or on an *ad hoc* basis).

Not only was the system of advance notification set up under the 1981 Wildlife and Countryside Act to be more permissive than the Porchester scheme (though it did apply to all protected land), it was also deemed to trigger a requirement for annual compensation payments calculated on a profits forgone basis. As a result it soon came to be seen as a treatment of the symptoms of the disease rather than a remedy of its cause and served merely to strengthen the hand of those conservation groups like the Council for the Protection of Rural England (CPRE) and Friends of the Earth (FoE) who were by now campaigning for agricultural policy reform (see FoE, 1983).

A key bone of contention was the compensation provisions themselves. Reflecting on the origins of these mechanisms, Whitby *et al.* (1990) come to the conclusion that they must be seen as products of an era of agricultural expansionism, invented as part of an expensive attempt to accommodate conservation to an ineluctable national policy. It rapidly became clear at the time that compensation would frequently be very substantial indeed and, moreover, would have to be financed by the NCC, national park authorities and local authorities rather than the Ministry of Agriculture. In a vivid demonstration of what happens when compartmentalized public policies work at cross purposes, large amounts of conservation money were now being expended to compensate farmers for the price support and capital subsidies that were forgone whenever a damaging operation did not go ahead. The verdict of many commentators was that by paying compensation within the existing system of agricultural protection 'conservation bodies are taking over some of the burden of agricultural support' (Bowers and Cheshire, 1983, p. 147). Its corollary was that agricultural support needed to be reformed, not only to remove the absurdities surrounding implementation of the Act, but also to improve the incentives for conservation in the wider countryside outside key sites. Sheail's (1995a, p. 86) comment was that:

> while the Government had [successfully] fended off demands for an extension of town and country planning to regulate farming operations, the political cost

was high. Not only might farming interests be resentful of any withholding of grant aid, but the avoidance by the Agriculture Departments of any liability for the costs of the consequent management agreements meant there was little sense of gratitude on the part of the conservation movement. There continued to be little sense of the Agriculture Departments becoming the driving force for consensus as to how the competing interests might be reconciled.

A debate about the future of the uplands, also coming to a head at this moment (CC, 1984), pointed strongly to the need for a reform of hill farming support if the farmed landscapes of the national parks were to be managed sustainably. Evidence of the way hill farming support had encouraged a steady intensification of upland land use over the preceding three decades was grist to the mill of reformers like MacEwen and MacEwen (1982), who developed an impressive environmental critique of Less Favoured Areas policy which revealed that, far from supporting the most disadvantaged farmers, the system of livestock payments and capital grants had actually favoured those larger, more intensively run farms best positioned to take advantage of subsidies calculated on a headage basis. This was having disastrous environmental and social effects, encouraging land improvement and overstocking and accelerating population decline. Hill farming policy, according to these critics, presented a classic case of the process through which large-scale capital for labour substitution contributes to demographic, social and environmental decline. It had to be reformed.

Proposals were brought forward by groups like the World Wildlife Fund (WWF; Potter, 1983) and the Council for National Parks (MacEwen and Sinclair, 1983) for reform of the CAP in order to switch part of the farm budget into conservation schemes. MacEwen and Sinclair's ambition was effectively to 'decouple' the payments offered to upland farmers from production by better calibrating headage payments to the degree of natural handicap and by imposing ceilings on the maximum amount of headage support individual farmers would be eligible to receive. In *New Life for the Hills*, they also proposed a system of Upland Management Grants which would pay farmers to manage the landscape resource. In their words:

> Conservation authorities are continually up against the fact that the right financial, fiscal or administrative tools have not been devised to promote the management of moorland, wetland, woodland, riverine, coastal or other habitats or landscape types, common land, or man-made features such as banks, hedges, walls and buildings. We think the time has come to move away from specific grants, or at the very least to complement them by a non-specific grant designed to promote both the initial investment and, where necessary, continuing management.
>
> (MacEwen and Sinclair, 1983, p. 38)

They were thus one of the first to work out what a system designed to pay farmers to produce countryside might look like in practical terms. The scheme put forward by this author (Potter, 1983) had more of a lowland focus and was particularly concerned with the problem of protecting farmed landscapes and habitat mosaics outside designated sites. The challenge here was to devise a system of payments which enrolled sufficient areas of land to safeguard and enhance larger tracts of countryside than would be possible through individually negotiated management agreements. The idea of giving farmers flat-rate payments to undertake countryside management was first hinted at by Porchester in his Exmoor report (Porchester, 1977). A little later, Feist (1978) had suggested that management agreements should be used more positively to promote good conservation management on farms. Now, under the proposed 'Alternative Package of Agricultural Subsidies', I was arguing for a tiered system of payments which would subsidize the retention of existing landscape and habitat features on farms and reward farmers who chose to undertake 'creative conservation'. All this would be funded by switching expenditure from production grants into conservation schemes. As O'Riordan (1984) commented, the aim was to 'set us off down a trail that ought to lead to the interlocking of agricultural and conservation expenditures around the notion of "sound husbandry".'

In 1984 an opportunity arose to test this latter idea in the field when the Countryside Commission initiated an experimental scheme in the Halvergate Marshes, a nationally important wetland landscape in the Norfolk Broads. A bitter dispute between the Broads Authority, a government agency responsible for the conservation of the area, and local farmers and land drainage interests keen to drain and plough the marsh and plant it to winter wheat, could not be resolved under the terms of the Wildlife and Countryside Act (for an analysis of the Halvergate Marshes saga, see O'Riordan, 1985). In effect, the Authority decided it could not and would not pay the high rates of compensation to desisting farmers which the legislation required (its calculations suggested that by the end of the decade over a third of the Authority's budget would be committed to paying farmers not to plant a subsidized crop that was already in surplus) and began looking elsewhere for a mechanism which would keep traditional livestock farming in place and so sustain the farmed landscape of the marsh as a whole. This was a rather revolutionary idea because it meant offering a payment to all farmers within the ring-fenced area, regardless of their intention to drain and plough. As a flat-rate system of payments, it would get away from the notion that farmers had to threaten to do something environmentally damaging before they would receive payment. Indeed, it would promote the idea that traditional farming was an activity for which the state was prepared to pay to see maintained. The CC's ingenious 'Broads Grazing Marsh Scheme', set up by the CC using experimental powers available under the Wildlife and Countryside Act, and funded initially by the Treasury, offered all farmers on the marsh a flat-rate annual livestock payment in return for an agreement to continue farming in a low intensity way. The effects were to be far-reaching. Without committing itself to the principle that livestock payments could be used for conservation ends (see below), the Ministry of Agriculture agreed not to offer conventional production grants to farmers in the area. The outlines of a partnership solution began to emerge whereby MAFF might be persuaded to take over subsidizing traditional livestock production in the Broads through the Grazing Marsh Scheme, while the Broads Authority took responsibility for shaping the management agreements concerned in order to meet its landscape protection objectives. The stage was set for further debate about the extent to which money from the farm support budget could be redirected into conservation schemes.

How much further these ideas and experiments would have been taken in the absence of other, politically more pressing reasons for farm policy reform is hard to say. These hints at the possibility of accommodation notwithstanding, there was little sign at this stage in either the US or the UK of the coalition building between the farm lobby and agri-environmentalists which Winters (1987) regards as a prerequisite for any sort of significant agricultural policy change. In philosophy and approach the two sides were still far apart. Commenting on the proposal for switching money into a new wave of conservation schemes in the UK, for instance, Elliot (1984, p. 26) asserted that:

> the NFU ... does not believe that it is the role of the agricultural budget or the MAFF to be responsible for mainstream conservation activities ... hence our belief that the measures in the Wildlife and Countryside Act provide for the foreseeable future a realistic means of tackling special conservation problems, where modern agriculture and conservation find it particularly difficult to accommodate each other. The current swingeing criticism of these measures is in our view extremely short sighted.

In America, the American Farm Bureau (AFB) was at best lukewarm about the idea of cross compliance, at worst openly hostile. Farmers could recall the last time a form of cross compliance had been in operation, when it required farmers receiving commodity support for one crop to comply with the set-aside requirements attached to other crops as well (Kramer and Batie, 1985). The arrangement proved very unpopular and the memory of it now coloured attitudes to the new conservation compliance proposals. Reformers were also struggling to catch the ear of policymakers; in Britain, the support of the Ministry of Agriculture for the Broads Grazing Marsh Scheme was strictly conditional and ministers continued to resist the implication that a more imaginative interpretation of the Less Favoured Areas Directive was all that was required to release money for environmentally friendly farming on a much broader front. In the US the ASCS conducted a vigorous campaign to derail the proposal for conservation compliance now circulating within SCS and in 1980 Congressman Foley, then chairman of the House Agriculture Committee, issued a press statement describing conservation compliance as 'coercive' and insisting that 'conservation programs should be based on co-operation with landowners in their voluntary efforts ... not the blackjacking of people into compliance' (quoted in Batie, 1985, p. 120). Less than a year later, however, a new US Secretary of Agriculture would be declaring his unequivocal support for conservation compliance while the British Ministry of Agriculture would be mounting an energetic campaign to introduce an agri-environmental component into an EC Regulation. What had changed?

The Budgetary Catalyst

To begin with, to differing degrees and for a variety of contextual reasons, policymakers in both the US and EU had come under intense pressure to tackle overproduction and reduce agricultural spending. This not only made officials more receptive to the multi-purpose agri-environmental schemes that were by now being put forward by reformers; it also facilitated a *rapprochement* between agri-environmentalists and a farm lobby now increasingly anxious to invent alternative, politically more defensible ways of supporting farmers' incomes. As Moyer and Josling (1990) have remarked, budgetary crises present the most potent, indeed probably the only truly effective, stimuli for sustained farm policy reform. The inertia of agricultural bureaucracy is so entrenched, and the identification with farming interests so strong, that an initiative for structural reform will only rarely emerge from within the policy machine.

> What makes a budgetary crisis so threatening ... is that it turns the policy process into a zero sum game where agriculture's demand for new resources is seen as coming at the expense of other programmes. No other crisis is likely to create the same kind of outside interest.
>
> (Moyer and Josling, 1990, p. 19)

Smith (1990) concurs, noting the way in which a budget crisis, by placing new constraints on policymakers, forces changes in attitudes and even agendas. Indeed, it can be argued that a policy as 'immunized' against reform as the CAP, is unusually dependent on external events to force anything approaching reform. A budgetary crisis creates a problem to which policymakers, sooner or later, have to respond.

The US farm crisis of the early 1980s had deep historical roots. Its symptoms were a lethal combination of falling export demand and rising real interest rates which translated into widespread illiquidity, increased surplus stocks and record federal expenditure on commodity support, but its cause was the over-investment in agricultural assets that had been encouraged over the preceding decade and a half by generous price guarantees, low interest rates and a buoyant export market for oilseeds and grains. Sheperd (1985) has documented a twofold increase in the debt of the US farm sector in the 10 years after 1972, reaching $215 billion by 1983. By the early 1980s annual interest expenses had reached $23 billion, averaging 50% or more of income from farm assets (Buttel, 1989). With a total capital stock valued at $1 trillion, the industry was no more highly leveraged in relation to total net worth than other industries but it was highly exposed as far as the ratio of interest payments to net income from farm assets was concerned. The problem was especially acute for larger businesses, up to 15% of farms with annual sales of more than $50,000 having debt asset ratios of more than 70% and thus being at risk of liquidation (Freshwater, 1989). The position of such farms was sustainable only as long as the appreciation in the value of farm assets kept up with this expansion in debt. In 1979 the Carter Administration broke the circle by tightening up monetary policy and raising interest rates in an effort to bring inflation in the wider economy under control. President Reagan intensified this policy and put further upward pressure on real interest rates by increasing the federal debt to finance military spending and tax cuts. At the same time a fall in agricultural export sales due to declining terms of trade and the effects of the debt crisis in developing countries put downward pressure on domestic prices and farm incomes. Reagan's restrictive monetary policy exacerbated this problem by strengthening the dollar on the international money markets. This appreciation of the dollar further choked off the demand for US farm exports. As Cochrane and Runge (1992, p. 54) observe:

> the Federal Reserve's decision to tame inflation not only drove down the value of holding real assets, such as farmland; it also contributed significantly to reduced export demand for what this land could produce. The growing integration of global capital and commodity markets made farmers hostages to exchange rate and interest rate policy as never before.

Farmland and asset values began spiralling downwards and many farmers were threatened with serious financial losses. Average land values declined by 19% between 1981 and 1985, with reductions of almost 40% in the hardest hit Mid Western states of Iowa, Illinois, Nebraska, Ohio and Minnesota (Buttel, 1989). These represented some of the deepest depreciations in land values since the Civil War (Helms, 1990). Expenditure on commodity support soared, rising from $4 billion in 1980 to over $18 billion during 1983/4 (Cochrane and Runge, 1992) as the federal government found itself obliged to put a floor under commodity prices and spend more on deficiency payments. Outlays on commodity price support would reach a record $26 billion in 1986. Although the Reagan Administration initially took a tough stance by threatening to make major inroads into public support, the farm debt crisis made this impossible politically. Brooks and Carter (1994) describe the formation of two competing schools of thought in the run-up to the 1985 Farm Bill – on the one hand, those arguing that US farmers should be allowed to export their way out of trouble, assisted by generous export subsidies and the end of hectarage controls; on the other, supporters of an expansion of supply controls in order to boost farm prices and raise farm incomes. With the Reagan Administration committed, at least at the level of rhetoric, to achieving steep reductions in these outlays through the forthcoming 1985 Farm Bill, and a farming community mostly anxious to maintain government financial support, the scene was now set for difficult negotiations over the future direction and content of farm policy.

European policymakers faced a crisis that was if anything even more securely rooted in the policy mistakes of the past. The budgetary implications of a high price, protectionist regime, maintained for 30 years in the face of mounting criticism, could no longer be finessed or ignored. As Mansholt had predicted, price support to maintain the incomes of marginal producers had over-rewarded the larger and more efficient farmers and led to the creation of surpluses. Once European Community self-sufficiency in temperate products had been reached and passed in the early 1970s the CAP could no longer be financed from levies imposed on imported products and the authorities had to intervene to maintain internal prices by removing excess production from the domestic market and subsidizing its storage or export (it would be another decade before it would follow the US in resorting to the expediency of a set-aside policy and even then only on a voluntary basis). Expenditure on the various market regimes consequently increased by more than four times between 1973 and 1985, eventually triggering the budget crisis which policymakers now faced. By the late 1980s, the EC would be devoting over 23 billion ECU, 70% of the entire EU budget, on market support. Dumping of cereal, beef, butter, milk powder, sugar and wine surpluses on international markets became common practice, depressing world prices and antagonizing trading partners (Gardner, 1996). Moreover, average farm incomes had languished despite a massive increase in government support. This was because farm costs had moved upwards to meet prices in an effort to choke off the stream of people and capital wishing to be part of a profitable industry (Harvey and Thomson, 1985). Howarth (1985, p. 88) comments that 'just as it is impossible to inflate one side of a balloon without inflating the other, so it is impossible artificially to raise the revenue of farmers without inflating their costs'. The result was a cost–price squeeze which narrowed net farm incomes at a time of record spending on price support, a paradox agricultural economists were only too ready to explain. As even the European Commission was by this stage admitting:

> a contrast between on the one hand such a rapidly growing budget and on the other agricultural incomes growing very slowly and an agricultural population in decline shows clearly that the mechanisms of the CAP as currently applied are no longer in a position to attain objectives laid down for agriculture in the Treaty of Rome.
>
> (CEC, 1985a, p. 65)

Later, it would elaborate the so-called 80/20 problem:

> Income support, which depends exclusively on price guarantees, is largely proportionate to the volume of production and therefore concentrates the greater

part of support on the largest and most intensive farms ... The effect of this is that 80 per cent of the support provided by the EU farm fund (EAGGF) is devoted to 20 per cent of farms which account also for the greater part of the land used in agriculture. The existing system does not take adequate account of the incomes of the vast majority of small and medium sized farms.

(CEC, 1991a, p. iii)

However, it was only when the budgetary limits on farm expenditure were in danger of actually being breached in 1983 that reform became a genuine possibility, the European Council meeting of that year demanding an examination of the CAP which would result in 'concrete steps compatible with market conditions being taken to ensure effective control of agricultural expenditure' (quoted in Tracy, 1989, p. 307). By European Community standards, this was strong language. With a fundamental imbalance between demand and supply looking set to continue for the foreseeable future, the Commission had already concluded that attempting to support farm incomes through open-ended price guarantees was no longer a sustainable policy option. As Tracy (1985) explains, special procedures were invoked, requiring the convening of a High Level Group of Agricultural Experts (the capitalization was significant), to bring forward proposals for reform. It took two EC meetings and a follow-up Agriculture Council session before a package of 'adaptations', which included the introduction of dairy quotas and a commitment to the future limitation of price guarantees, was finally agreed on 31 March 1984. The Commission maintained the pressure for reform by publishing the so-called 'Green Book' in the following year (CEC, 1985a), setting out a programme for longer term reform which began from the recognition that the era of open-ended price guarantees was at an end.

Argument and Persuasion

For agri-environmental reformers in both the US and the UK these were encouraging developments, opening up a larger debate about farm policy reform and creating the opportunity to make a link between the financial bankruptcy of agricultural support and its environmental effects. In the US, lobbyists had the additional advantage of a legislative process which every 5 years brought farm policy matters to a head in the form of a federal farm bill. The 1985 Farm Bill, however, promised to be particularly hard fought, with the Administration introducing a radical draft bill into the House of Representatives in early 1985 designed to 'get government out of agriculture'. Although subsequently defeated in the Senate, this zero option set a benchmark against which versions of the bill now being developed by the House and Senate Agriculture Committees could be compared. In the event, these bills preserved intact most of the existing commodity programme mechanisms and there was a strong commitment to continued support of farmers' incomes through price support. The Act that was finally passed in Congress expanded annual hectarage reduction programmes and boosted export subsidies, particularly on exports to European markets. But budgetary constraints could not be ignored and policymakers searched for mechanisms to include in the bill which would limit budget exposure and keep domestic output under control. It was into this fertile ground that the American Farmland Trust (AFT), now emerging as the leader of a new coalition of conservation organizations, sowed the idea of a Conservation Reserve Program designed to take cropland out of production and save soil (AFT, 1984). Its influential *Soil Conservation in America. What Do We Have to Lose?* asserted that:

> the public is growing impatient with the growing impasse in the struggle against erosion problems, particularly with government processes that work at cross purposes, some promoting conservation while others discourage it ... [On the other hand] we have barely begun to use what we know about soil erosion and conservation needs to formulate responsible public policies in the United States.
>
> (AFT, 1984, p. xxii)

The CRP fulfilled this last requirement by targeting federal resources at highly erodible land, a new category of farmland identified from data recently available under the NRI. It had been estimated that just 6% of total cropland erosion (10 Mha) accounted for 43% of the total tonnage of sheet and rill erosion (811 million tonnes per year). The AFT's review of research on the productivity and off-site impact of erosion suggested that most of it emanated from locations where soil loss exceeded natural restoration by a factor of three or more. Under the CRP, farmers would be offered 10-year contracts to take this vulnerable cropland out of production and plant it to grass or trees.

The idea was not new, but its presentation as a multi-purpose instrument by policy entrepreneurs like Cook (1989) and Ogg *et al.* (1984) was, and AFT took pains to stress its efficiency as a conservation policy which protected the soil and improved programme consistency but also preserved farmers' right to choose (AFT, 1984). As the AFT report points out:

> At present, USDA production adjustment measures have (two) serious shortcomings with respect to achieving conservation. First, the programs operate on a year to year basis, making it virtually impossible for farmers to incorporate short-term program requirements into a long-term conservation system for their land. The annual process limits soil savings at the same time it encourages cultivation of highly erodible soils to preserve the farm's base hectarage ... Second, no attempt is now made to selectively enrol land into the diversions and set aside. Frequently, farmers idle low, wet land or land susceptible to droughts, which not only compromises the production adjustment goals of the programs, but also further reduces the program's soil conservation benefits. The CRP would be long term, and would be aimed only at highly erodible land.
>
> (AFT, 1984, p. 99)

The support of farm groups like the National Farmers' Union (NFU) and the National Farmers' Organization (NFO) was consequently relatively easily secured. As Browne (1988) observes, these groups let it be known that the Reserve idea would be useful as a cost-cutting mechanism which, by taking sufficient cropland out of production (the AFT was arguing for a target enrolment of over 16 Mha at this stage), made it easier to justify continued high price support.

After some initial reluctance, policymakers liked it too, and for much the same reason, and in June 1985 Secretary Block announced his support for an 8 Mha Conservation Reserve modelled largely along AFT lines. By this stage, conservation compliance was also enjoying a fair wind amongst legislators, John Block having already declared himself in favour of cross compliance as a low cost way of promoting soil conservation. At a time of heightened sensitivity to charges of inefficient government, the Secretary was keen to rebut charges that his department was simultaneously penalizing and promoting soil conservation on the nation's farms. His declaration followed swiftly on the heels of an expression of support from both the House and Senate Agriculture Committees for the idea that commodity support should be conditional on conservation conditions being met – or even be withheld where sod- or swamp-busting occurred (Batie, 1985). This was surprising given the more coercive nature of the proposal (USDA continued to dislike it and would later produce research which questioned its likely effectiveness: see Reichelderfer (1985) and the discussion in Chapter 3 below), but is probably explicable in terms of the political capital to be gained by being seen to be reducing inconsistencies between the commodity and the conservation programmes. Commentators like Miranowski and Reichelderfer (1985) had done much to obtain recognition of inconsistency as a problem which needed a solution, documenting the increasing divergence between soil conservation and commodity programme goals. According to these analysts, a plot of land first brought under the plough through subsidies allocated under the Homestead Act of the late 1800s could easily have been planted back to grass under the New Deal programmes, ploughed up again in the 1970s and then regrassed under a conservation contract with the USDA. 'No wonder the public is raising questions regarding the consistency (of public policy)' (Miranowski and Reichelderfer, 1985, p. 209). A Harris public opinion poll taken in 1979 for the RCA found that over 75% of the sample agreed with the statement that federal farm commodity support should be withdrawn from farmers who did not conserve soil and water. A later poll of farmers commissioned by USDA as part of the RCA process discovered that while only 49% of the sample directly favoured cross compliance, over 60% believed it made good policy sense and would come about. In addition, there was always the possibility that, by removing non-complying farmers from farm programmes conservation compliance would make some contribution to reducing the aggregate costs of support (Ervin, 1985). For the farm lobby the political advantages of conservation compliance were already clear, for here was a way of making income support to farmers more publicly defensible by delivering an environmental improvement in return for payments received. Farmer support for the concept had first emerged from the RCA, where it came to seen as a way of defusing criticism and avoiding the imposition of compulsory controls. Indeed it was the Colorado Cattlemen's Association and the Colorado Farm Bureau Federation which, convinced of the efficacy of compliance, first campaigned for the 1984 sodbuster legislation (Kramer and Batie, 1985). Now a larger calculation was being made by national representatives of its defensive value as a permanent feature of farm programmes. By the autumn of 1984 the Ninety-ninth Congress had before it an unprecedented number and variety of proposals for agri-environmental reform, supported by a powerful coalition of interests. The scene was set for the most environmentally productive piece of farm legislation since the New Deal.

In a more oblique way, the debate about reform of the CAP during the mid-1980s intersected with agri-environmental concerns in the UK. There was not to be anything as dramatic as a single, all-embracing farm bill but there was a gathering sense of the inevitability of reform and this made farm groups increasingly keen to treat with agri-environmentalists and their various policy proposals. Publication of the EC's Green Book in 1985 opened up the debate about CAP reform, with the Commission canvassing a long-term strategy designed to close the gap between Community and world prices (CEC, 1985a). The report makes clear the Commission's preference for price adjustments rather than supply control measures such as quotas and puts a strong case for replacing price support with direct income aids to vulnerable or deserving farmers. A little earlier, the President of the Court of Auditors, Pierre Lelong, had proposed a 40% cut in grain price support under the CAP in order to liberalize agricultural markets and defuse a looming CAP-related trade dispute with the US (Lelong (1983), quoted in Duchêne *et al.* (1985)). As even the EC was admitting 'If Community agriculture is to succeed – as it should – in expanding its exports and maintaining its share of world markets, it must increasingly accept the market disciplines to which other sectors of the Community's economy are subject' (CEC, 1983, p. 7). This possibility suited agri-environmentalists well, up to a point. Cutting output prices and 'decoupling' support promised some lessening of the economic incentives to intensify production and it was argued by many commentators that a sustained reduction in price guarantees should bring about an environmentally beneficial extensification of production on a broad front (Jenkins, 1990). At a landmark seminar convened by the CPRE, WWF and the Institute for European Environmental Policy (IEEP) in March 1985 (Baldock and Conder, 1985), there was wide agreement that cutting CAP prices was a desirable element of any reform package, but that this should be accompanied by changes in structural policy designed to protect farm incomes in environmentally useful ways – 'direct environmental remuneration' as Heino von Meyer, a keynote speaker at the seminar, put it. This last proposal was well received by the farm lobby, by now increasingly aware of the threat to their members' historical policy entitlements of any move towards a more decoupled CAP. By making any government transfers to the farming community much more visible and transparent, decoupling threatened to open up a debate about the legitimacy of supporting farmers because they are farmers. The lobby would now become increasingly receptive to the idea of green payments and, later on, a version of American-style conservation compliance, because these appeared to offer a publicly defensible way of continuing with agricultural support. The first public evidence that this calculation had been made came in evidence to a House of Lords Select Committee Inquiry into agriculture and the environment held in the autumn of 1984 (House of Lords, 1984a).

The Inquiry, ostensibly set up to advise the government on a suitable response to the recently published EC proposals for a new Regulation on Agricultural Structures, was an important event, allowing parties to parade their proposals and define the substantial consensus about the need for reform which by then existed. The main bone of contention for environmentalists was MAFF's narrow interpretation of the existing Less Favoured Areas Directive (a central component of Agricultural Structures policy) and its refusal to countenance any deployment of funds under the Directive for environmental purposes. According to the government's own conservation agencies, there was a provision under Article 3(5) of the original Directive which allowed member states to offer direct payments to farmers 'disadvantaged' by environmental regulations such as might apply within an SSSI or a national park (CC, 1984). The experience of The Netherlands was presented, where 'permanent natural handicap' had been interpreted for some years on environmental as well as physical land capability grounds. Here, farmers who found themselves in areas of high landscape value, and thus subject to restrictions on farming practice emanating from environmental legislation, were deemed to qualify for payment under the Directive. As the White Paper on the Relation between Agriculture and Nature and Landscape Conservation saw it, the LFA Directive was an important plank in the government's policies for realizing the objectives of nature and landscape management then being developed in The Netherlands (Bentham, 1984). Witnesses, including those representing the Country Landowners' Association (CLA), contended that a similarly liberal definition of handicap could be exploited by agriculture departments in the UK to fund measures such as the Broads Grazing Marsh Scheme (BGMS), then still being financed directly by the Treasury. The CLA quoted a Commission statement, issued in response to a European Parliamentary question about the Directive, that, 'while it may not be used to encourage conservation *per se*, it is to be used for the encouragement of farming which, in turn, will have a positive effect on the countryside'. Few witnesses demurred from the CC's assertion that, here, and elsewhere, 'agricultural policy should discriminate in favour of activities which benefit the environment and against those which are detrimental to it' (House of Lords, 1984a, p. 163).

MAFF refused to concede this key argument by noting that the legislation permitted natural handicap to be defined only in physical, agricultural terms; moreover, environmental protection could not be funded under the new regulation because it derived its legal origins from the Treaty of Rome, which specifically refers only to agricultural policy goals. The Committee roundly criticized MAFF (and, by association, the Department of the Environment (DOE)), for taking such a pusillanimous stance, remarking that:

> this attitude fails to take account of the possibility of recasting present expenditure to give greater emphasis to environmental measures. It also appears to overlook the possibility that, by making a more liberal use of existing provisions to include environmental objectives, large amounts of Community funds could be obtained.
>
> (House of Lords, 1984a, p. xviii)

A sister report took both the MAFF and DOE to task for their poor coordination of research and development relating to environmentally friendly farming and recommended a fundamental rebalancing of the agricultural research spend (House of Lords, 1984b). By now the CLA had joined forces with CPRE in calling for 'changes in policy which would help end the unfortunate and damaging conflicts over the effects of modern agricultural practices on the countryside' (quoted in Lowe *et al.*, 1986, p. 182). In a joint statement issued on the eve of the House of Lords debate on the Select Committee's report, the two organizations committed themselves to 'working urgently together to obtain adjustments to agricultural support policies, which will establish a better balance between efficient agriculture, private land ownership and the public interest in conservation and enjoyment of the countryside'. Replying to the report on behalf of the government, Lord Belsted, Minister of State at MAFF, restated the official line on Article 3(5), but then announced an intention to seek a completely new title to the Structures Regulation in order to 'enable us in environmentally sensitive areas to encourage farming practices which are consonant with conservation'. The first reaction of commentators was to entertain doubts about the Ministry's motives in taking such an apparently bold step (Lowe *et al.*, 1986). In retrospect, the announcement was a watershed event, marking the first stage in MAFF's conversion to the idea of greening the CAP. The success of its campaign in Brussels would pave the way for all the agri-environmental policy developments to follow.

Agri-environmentalists would probably have found it harder to persuade the agricultural policy community of the case for agri-environmental reform without the budgetary crises of the mid-1980s. On the other hand, by describing the way lobbyists, commentators and policy entrepreneurs first critiqued farm policy and then brought forward proposals to reform it, this chapter has also emphasized the importance of 'discourse, argument and persuasion' in the long policy process. In both the US and UK, reformers had first to gain acceptance of what has been called here 'the policy thesis' – the notion that agricultural support has been a driver of environmental change in rural areas – before they could begin to lobby for policy reform. In the US they faced the additional complication of explaining why existing federal soil conservation programmes were failing to meet the need for more soil conservation on farms. Much of the new thinking was opportunistic, of course, policy entrepreneurs in both countries developing packages of green reforms which addressed over-production and the problem of farm incomes as well as environmental decline. Lobbyists exploited the agricultural policy situation in which they found themselves, developing programmes which would attract support from within the agricultural policy community as well as from among environmentalists. What is nevertheless impressive from this distance is the speed with which relatively sophisticated policy ideas like the use of environmental contracts, conservation compliance and the CRP were introduced into the debate at this critical moment. It would not be very long before they would be enshrined in legislation.

3 The Conservation New Deal

President Reagan signed the Conservation Title of the 1985 Farm Security Act into law on 23 December 1985. According to some, it was the most progressive piece of conservation legislation since the Depression. The World Watch Institute, a touch hyperbolically perhaps, rated it as one of the three greatest US environmental policy achievements of the twentieth century. Kenneth Cook, a prime mover behind the Title, argued more realistically that it marked the beginning of a new social contract between farmers and government which would eventually transform the nature of agricultural support. Even Peter Myers, Deputy Secretary at USDA, was willing to admit that the Title 'wove conservation into the very fabric of farm policy' for the first time (Myers, 1988, p. 10) and was a fitting climax to the campaign to green agricultural support. In the event, the Title's passage through Congress was remarkably smooth and uncontroversial. A tight coalition of environmental groups encountered virtually no opposition to its proposals. Indeed, the House and Senate Agriculture Committees actually competed to put forward the 'greenest' bill (Zinn, 1991). The attention of those who might have been expected to derail the conservation proposals – chiefly the coalition of commodity groups – was diverted by a fierce battle then under way about the degree of commodity support that should be written into the main part of the Bill. Anyway, environmental reform was by then widely seen as irresistible and desirable by both policymakers and their farming constituencies. As Swanson (1993) observes, what makes the Conservation Title so interesting is the diversity of the political coalition that produced it. The environmentalist case for reform had been well made and skilfully presented but the calculation that changed minds was that environmental reform could be a 'magic bullet' which would save soil, cut production, and support farmers' incomes. Batie (1986, p. 4) comments that 'while there are inescapable problems of choice and trade-off with these three

Farm Bill goals ... as objectives (they) reinforced one another to the extent that supporters of each objective were willing to help each other pass the Bill'. The result was a true policy omnibus on to which all interested parties could climb.

The 1985 Conservation Watershed

Proposals for a Conservation Title featuring a Conservation Reserve and versions of conservation compliance were put forward early in 1985 by the Administration and by the House and Senate Agriculture Committees (Malone, 1986). These differed only on points of detail (the Senate Bill envisaged a Reserve of 12 Mha compared with the House's 10 million) and were quickly reconciled in a joint conference held in December. Perhaps the most surprising proposal at this stage came from the Senate Agriculture Committee, which suggested that all federal assistance should be denied to farmers who continued to cultivate erodible land after 1988. The draft bill which emerged from this conference, incorporating a version of this 'sodbuster' provision together with a proposal for a 14 Mha Conservation Reserve, underwent no further revisions and became law soon after. Its centrepiece, and by far the most expensive element in the Title, was to be the Conservation Reserve Program (CRP). Under Section 2 of the Title, the Secretary of Agriculture is required to 'bring about a CRP ... to assist owners and occupiers of highly erodible cropland in conserving and improving the soil and water resources of their farms and ranches'. Ten-year contracts are to be offered through a system of competitive bidding in order to pay for a diversion of cropland into grass or trees. The Act envisages that between 16 and 18 Mha will be enrolled in these contracts between 1985 and 1990 and specified that one-eighth of the Reserve should be planted to trees. To qualify, cropland must be classified as 'highly erodible'. Specifically, at least two-thirds of a field must be eroding at twice the tolerance or 'T' rate – the accepted national average at which experts believe soil loss can be tolerated for various soil types without endangering long-term soil productivity (usually 1.98 t ha^{-1} per year or less). At the outset, USDA estimated that 42 Mha would be eligible for the Reserve on this basis (Myers, 1988). Meanwhile, under the accompanying 'Highly Erodible Lands' (HEL) sub-title, various checks would be introduced in order to buttress the CRP and 'discourage the cultivation of highly erodible land and avoid production from land that would otherwise be cultivated'. The conservation compliance provision empowers the Secretary to require all farmers with highly erodible land to draw up and implement a conservation plan for their farms in order to continue to qualify for USDA programme benefits. This plan was to be fully implemented by 1995. Should any farmer plough up grassland after 23 December 1985 where there is an erosion hazard but no conservation plan, he or she would be deemed to have 'sodbusted' and under the sodbuster provision could be denied USDA benefits. Equally, under the swampbuster provision, farmers who drain and convert wetlands and plant them to crops will be barred from receiving commodity support, disaster payments or crop insurance. Analysts at USDA estimated that 48 Mha of cropland would fall under conservation compliance (Osborn and Miranowski, 1994).

The seemingly rather draconian nature of the HEL sub-title surprised some commentators at the time. Whereas the CRP was full square in the tradition of 'uncoerced voluntarism' that had characterized federal soil conservation programmes since the 1930s, here was a development which appeared to erode farmers' policy entitlements in a very significant way. In fact, conservation compliance predates the CRP as a policy idea and was attracting broad bipartisan support as early as 1981. Congress had considered sodbuster legislation in the previous two sessions and both chambers had passed provisions during the 98th Congress. The first sustained campaign for a form of conservation compliance had been mounted in the Great Plains during the late 1970s, where a large-scale conversion of grassland to wheat had exacerbated a wind erosion hazard and blown large quantities of soil over irrigated pasture and along fence lines (Helms, 1990). Colorado Senator William Armstrong had been persuaded by affected landowners to introduce a bill that would deny USDA programme benefits to farmers who converted fragile land to crops. This had attracted wide support, including the first tentative expressions of interest from the NACD (Kramer and Batie, 1985). Armstrong then introduced an amendment to an agricultural appropriations Act entitled 'Prohibition of Incentive Payments for Crops Produced on Highly Erodible Land' and this was still being debated when the 1984 Farm Bill first surfaced. So far as the 1985 Conservation Title was concerned, there is as much in the argument that the CRP was necessary to sugar the conservation compliance pill which politicians by this stage had decided farmers needed to swallow, as that compliance was an afterthought, albeit a rather radical one. Ervin *et al.* (1991) comment that, while the CRP and compliance were seen as mutually exclusive alternatives at the beginning of the Farm Bill debate, they were eventually regarded as mutually supportive and were deliberately woven together into the fabric of the new Title.

There is evidence, in fact, that by combining a targeted CRP implemented through competitive bidding with the stick of conservation compliance and sod- and swampbuster, legislators were actually trying to optimize the goals of soil conservation, supply control and reduced government spending. Research undertaken by the Economic Research Service (ERS) of USDA (Ogg *et al.*, 1984) had already established that competitive bidding saves money by encouraging farmers to offer land for a price which reflects the opportunity cost of enrolment. Specifically, bidding eliminates the windfall gains which tend to be associated with a flat-rate 'offer' system (under which farmers with land of different productivity receive the same amount of compensation) and matches payments to what individual pieces of land would be capable of earning if they had remained in an agricultural use. The ERS had developed economic models which persuaded USDA managers that significant budgetary savings were possible, provided a bid acceptance procedure was adopted which favoured bids from farmers offering the greatest erosion reduction for every dollar spent (GAO, 1989). Applied to the 1978 land retirement programmes, in fact, it was estimated that overall costs could have been cut by up to 25% had competitive bidding succeeded in eliminating windfall gains. The thinking behind conservation compliance was that, by requiring all farmers to submit, implement and substantially self-fund a conservation plan for erodible cropland, compliance would propel more farmers into the voluntary CRP and make it more likely that they would submit competitive bids. Competitive bidding would achieve more conservation for every dollar spent, while sodbuster and the installation of conservation plans would anticipate the moment when CRP contracts expired, building in safeguards against an indiscriminate 'ploughout' of CRP land where this was likely to be environmentally damaging. At the same time, and this was the big consideration for USDA, by displacing the farmer's 'base acreage' on which crops were planted and commodity support received – what Taft and Runge (1987) call the 'base bite' effect – land enrolled in the CRP was supposed to reduce production and hence government spending on deficiency payments. In addition, the reduction in output from participating farms was expected to push up commodity prices and thus lower the deficiency payment that had to be paid on remaining crop production. Further budget savings would be made. According to USDA's initial assessment, almost 75% of the land eligible for the CRP was planted under programme crops, 50% of which was under corn and wheat. Its projections suggested that the CRP would pay for itself. (Though after 1991 the ERS revised this to predict that savings from the annual commodity programmes would more than cancel out CRP costs. Over the life of the programme, however, ERS estimated that direct CRP costs would be $2–6 billion greater than any reduction in expenditure on deficiency payment because of the 'base bite' effect.) According to estimates produced by the AFT, on the other hand, a 14 Mha CRP could result in a net (undiscounted) saving on annual price and income support of over $0.6 billion for the crop years 1986–1990 (AFT, 1984).

Enrolling Land, Saving Soil

The ASCS moved with almost unseemly haste to bring the new Conservation Reserve into existence. During the first three sign-ups over 3.3 Mha were enrolled. By the winter of 1987 a further 4.3 Mha had been drawn in after farmers had been offered a one-time bonus. Farmers clearly found attractive the idea of obtaining a guaranteed government cheque in return for agreeing to retire land. Participation in the CRP was judged to be potentially profitable and to entail little risk; by comparison, investing in soil conservation practices without substantial subsidy from government was rarely without risk and could be costly. According to early studies of the farmer response, many participants were already aware of the need for more soil conservation on their farms, non-adopters being the people who denied there was a problem or who lacked information about the programme (Esseks and Kraft, 1986). Few studies found much self-selection taking place among participants, with all sizes and types of farms represented (see, for instance, Lovejoy and Napier, 1986). This was consistent with what Batie (1988) calls 'the traditional model' of farmer decision making on which the CRP had been based and which can be summed up in the phrase: 'more conservation subsidies buy more conservation'. By 1990 the Reserve stood at 13 Mha (Ervin *et al.*, 1991). After a further three sign-ups (see Table 3.1), the Reserve peaked in 1996 at 14.7 Mha or 8% of total US cropland – an area roughly the size of the state of Iowa and accounting for a significant proportion of the total area of land set aside in the US overall (see Table 3.2). A total of 375,000 contracts were in force (Osborn, 1993). By this time annual expenditure totalled $1.8 billion or 9% of the $131 billion expended by USDA on commodity support in that year. The first soil conservation assessments of the programme suggested an impact on soil erosion commensurate with these huge enrolments and a decade after it was commissioned the CRP was credited with reducing erosion in the US by 670 million tonnes or 25% of the total each year (Margheim, 1994). Of all US cropland thought to be highly erodible, 27% had been enrolled in the Reserve by early 1993 (Dicks, 1994). An estimated 88 million tonnes of sediment has been kept out of waterways during each year of the programme's operation. Over 1 Mha or 10,000 km^2 had been planted to trees and the Reserve has

Table 3.1. Land enrolled in the US Conservation Reserve Program. Source: based on data in USDA (1993).

	Sign-up period	Number of contracts (in thousands)	Cumulative total area enrolled (Mha)	Cumulative percentage enrolment
1	March 1986	9.4	0.30	2.0
2	May 1986	21.5	1.42	9.6
3	August 1986	34.0	3.32	22.6
4	February 1987	88.0	7.16	48.6
5	July 1987	43.7	8.96	60.9
6	February 1988	42.7	10.33	70.2
7	July 1988	30.4	11.38	77.3
8	February 1989	28.8	12.37	84.0
9	August 1989	34.8	13.72	93.2
10	May 1991	8.6	13.91	94.5
11	July 1991	14.7	14.31	97.2
12	June 1992	18.4	14.72	100.0
Total		375	14.72	100.0

Table 3.2. Cropland idled (in millions of hectares) under US federal farm programmes, 1984–1995. Source: USDA (1993) and OECD (1997).

Year	Annual programmes	CRP (cumulative total)
1984	10.9	0.30
1985	12.5	1.42
1986	18.6	3.32
1987	24.3	7.16
1988	21.4	8.96
1989	12.5	10.33
1990	11.3	11.38
1991	12.1	12.37
1992	8.1	13.72
1993	9.3	13.91
1994	5.3	14.31
1995	5.7	14.72

engineered a substantial expansion in the area of wildlife habitat on farms (Berner, 1994). Estimates of the total monetized environmental benefits of the CRP (including gains in soil productivity, improved water quality and expanded wildlife habitat) have been put at between $6 and $13 billion over the life of the programme (Crosson, 1996). Another estimate made by Alexander (1989) suggests that annual damage costs from sediment in the south had declined from $643 million to $44 million after the first six sign-ups. Significantly perhaps, given the greater restrictions it enshrined, the HEL sub-title had a much slower start, waiting in the wings until June 1987. Even then, implementation of cross compliance was delayed by arguments about the criteria on which compliance would be judged. Nevertheless, by 1988 SCS had managed to draw up 65% of plans affecting over 36 Mha of cropland and by 1995 had brought nearly all eligible cropland into a conservation plan of some description. It is estimated by Ervin (1993) that 16 Mha of conservation tillage, 4 Mha of contouring, 350,000 km of terraces and 0.5 Mha of grassed waterways had been or were planned to be installed under these plans. Dickason and Magleby (1993) calculate that the resulting erosion saving would be over 372 million tonnes per year or 12% of the total. Combined with the effects of the CRP, they estimate that the 1985 Conservation Title may have achieved more than twice the amount of erosion control accomplished under all previous soil conservation programmes operating since the 1930s.

Despite these achievements, by the end of the 1980s the CRP was being cast in an increasingly critical light as it became clear how much more conservation might have been bought for the large amounts of public money actually spent. Critics pointed to the poor targeting of erodible land and to USDA's failure to operate the system of competitive bidding as intended. It was muttered that the CRP, as implemented by ASCS, was much more efficient at supporting farmers' incomes than conserving soil. Moreover, the use of aggregated estimates of the environmental impact of the programme almost certainly exaggerated its true effects (Dicks, 1994). Many studies measured the movement of soil within a field, not its movement off the farm. Thus, estimates of the tonnage of soil 'saved' in official studies did not necessarily translate into an equivalent improvement in environmental quality. In addition, studies made no allowance for changes in technology. The significant expansion in conservation or no-till since the CRP was introduced meant that at least a proportion of the improvements in soil conservation attributed to the policy may actually have been due to background changes in farming practice. The political economy of agri-environmental reform was undergoing a slow shift as the nature of the trade-offs that had been made by USDA between conservation, income support and supply control became clear and the realization dawned on agri-environmentalists that, to stand any chance of being recommissioned, the CRP would have to demonstrate much better environmental value for money. The CRP was undoubtedly expensive. After 12 sign-ups, the policy had absorbed $11.6 billion or 9% of the $131 billion spent on commodity support since 1986 (Heimlich and Osborn, 1994). CRP costs were 50% of the $23.3 billion USDA had invested in all soil conservation programmes and initiatives since this date. The extent of the environmental return on this investment was now the subject of protracted debate.

The problem was the substantial degree of discretion that had been granted to USDA and its agencies in implementing the Conservation Title and the resulting gap between the legislation and its implementation. According to Dicks and Grano (1988), this is not unusual where broad, multi-purpose legislation is concerned but in the case of CRP, USDA's discretionary powers were particularly strong. In effect, the Secretary of Agriculture could determine who would participate in the programme (eligibility rules), what level of participation would be allowed (enrolment targets) and how participants were to be selected (bid acceptance criteria) – the critical variables in determining how much land is reserved and where it is located (Reichelderfer and Boggess, 1988). Initially, eligibility conditions were strictly defined to target highly erodible land; indeed, prior to the second sign-up the definition was actually tightened up in response to conservationists' demands to narrow the focus to land eroding at 3T (three times the 'tolerance rate' judged to pose no long-term threat to agricultural productivity), a decision which reduced the area of eligible land from 42 to 28 Mha and appeared to underline USDA's commitment to putting conservation first (Dicks et al., 1987). After the fourth sign-up, however, this was relaxed to embrace the 49 Mha of cropland thought to be eroding at T as it became clear that ASCS's aim was to maximize enrolment for a given budgetary outlay. Although justified at the time on the grounds of consistency (land eligible for CRP was now the same as that eligible for conservation compliance), the effect of this rebalancing was to level down rather than level up definitions and bring about a significant dilution of the programme's conservation mission. (Dicks and Vertrees (1987) point out however, that defining the eligible hectarage in relatively liberal terms should in principle give policymakers greater flexibility in targeting other objectives. The larger the eligible area, the greater the likelihood of receiving more bids to retire land with a broad range of environmental characteristics and potentialities.) Moreover, because the terms of conservation compliance were themselves amended – the SCS deciding in 1988 under pressure from the farming community to insist only on the adoption of least-cost conservation practices under an 'Alternative Conservation System' rather than a strict reduction in rates of soil erosion to the T standard – farmers faced less pressure to enrol land in the Reserve; or if they did decide to enrol, would require a larger financial inducement to do so. Again, the Secretary of Agriculture was given significant powers of discretion in determining the rigor with which compliance would be pursued. The original reconciliation conference had conceded that:

> If a rigid standard of 'T' value is mandated for an acceptable conservation plan, even if erosion has been reduced from, say, 12 tonnes per hectare to 3 tonnes per hectare through the application of cost-effective conservation measures, the producer could be required to either install a very expensive additional practice such as terraces or convert the land to grass or trees from cropland in order to continue to be eligible for programme benefits. It is not the intent of the Conferees to cause undue hardship on producers to comply with the provisions. Therefore, the Secretary should apply standards of reasonable judgement of local soil conservation professionals and consider the economic consequences in establishing requirements.
>
> (Quoted in Helms, 1990, p. 40)

SCS's decision to relax the compliance standard was seen as perfectly consistent with this philosophy. Unfortunately it had the effect of threatening to negate one of the most important synergies built into the Conservation Title.

It fell to the General Accounting Office (GAO), in a report published on the CRP in 1989, to take the first brick out of USDA's conservation wall (GAO, 1989). Entitled *Conservation Reserve Could be Less Costly and More Effective*, its central message was that, while the conservation impact of the CRP was significant, more could have been achieved for a smaller outlay of public money if the Department had better targeted the programme and operated the bidding system as intended. In effect, USDA had sacrificed environmental benefits, particularly from improved water quality, in order to meet hectarage enrolment and tree planting goals. Regarding the planting of trees on CRP land, USDA decided to relax the eligibility criteria for farmers wishing to afforest their land after discovering that tree planting under the early sign-ups was severely under target. Environmentally speaking, this was superficially a welcome adjustment: experience with the Soil Bank shows that a tree cover is much less likely to be reversed than one of grass, 80% of all land planted to trees under this programme still being forest 35 years later (Brandow, 1977). As GAO now pointed out, however, USDA had effectively traded soil conservation and water quality against this objective by allowing only moderately erodible land to enter the CRP. 'On balance, USDA's decision to relax the eligibility criteria to encourage tree planting has detracted from the overall effectiveness of the programme in meeting its full range of objectives – particularly in the areas of soil erosion and sedimentation' (GAO, 1989, p. 28). GAO went on to note that the original legislation gave the Secretary discretion to include in the programme 'lands that are highly erodible but that pose an off-farm environmental threat' but observed that up to 1988 this had not been activated, despite a growing public recognition that water pollution rather than productivity loss was overwhelmingly the most important environmental cost of soil erosion. Rather, the Department had stuck to its traditional concern with soil erosion as an on-farm problem, favouring through its eligibility rules and acceptance procedures land threatened by erosion from wind in the Great Plains over areas where erosion due to water is most likely to lead to surface water pollution. In the main:

> while CRP benefits are substantial, the overall impact and effectiveness of the programme could have been enhanced if USDA had managed the programme to address the full range of CRP objectives instead of focusing on the need to enrol prescribed hectarage amounts. As a result, the soil savings decreased and other benefits, like reduced sedimentation and improved water quality, were not attained.
>
> (GAO, 1989, p. 3)

What seems to have happened is that the tendering system was operated to ensure that more generous annual payments were offered in the Mountain and Plain States in order to enrol most land from these locations. At the same time market prices being received for the corn and wheat predominantly grown here were low relative to CRP payments during these early sign-ups, further encouraging farmers to put land into the Reserve. The result was a disproportionate enrolment of land in the Great Plains and Mountain States relative to water-erodible land in the Corn Belt, Lake States, north-east and south-east (Ervin, 1988). In fact, as Table 3.3 shows, 61% of these early CRP enrolments came from the Northern and Southern Plains and the Mountain States despite only 46% of total eligible land available for the CRP being found in these states; 50.6% of the total CRP erosion reduction was achieved through a reduction in wind erosion specifically (Osborn, 1994). On the basis of Ervin's (1990) analysis, this is a far from ideal pattern of enrolment if the aim is to maximize environmental benefits. Comparing a number of different 'implementation pathways', Ervin demonstrates that the strategy actually followed was the one best calculated to minimize average rental and cost-share outlays. In order to maximize the environmental benefit on every dollar spent, USDA should have targeted the north-west and Lake States and the Delta and

Table 3.3. Distribution of CRP enrolments, 1985–1988. Source: GAO (1989).

Region	Percentage of soil loss on eligible area due to:			% of area enrolled in CRP	% of soil saved on enrolled area
	Wind	Water	Total		
Mountain	13	3	16	20	19
Northern Plains	5	7	12	25	20
Southern Plains	22	3	25	16	26
Subtotal	40	13	53	61	65
North-east		3	3	1	–
Lake	1	3	4	8	7
Corn Belt	1	25	26	14	13
Appalachian		7	7	3	4
South-east		2	2	5	4
Delta		2	2	3	3
Pacific	1	2	3	5	3
Subtotal	3	44	47	39	34
Total	43	57	100	100	100

Appalachian regions. Instead, it had drawn in disproportionate amounts of cropland from areas where the environmental hazard of keeping land in crop production was relatively low. Pursuing the same theme, Reichelderfer and Boggess (1988, p. 9) analysed bid acceptance procedures and the way bid pools were defined to conclude that 'net government costs could have been reduced while simultaneously increasing the extent to which erosion and supply control objectives were met'.

Reflecting on these findings, it was the perceived disproportionate expense of the CRP relative to the benefits actually being generated which now began to attract critical comment. As has been said, the assumption made by those who originally designed the CRP was that it would save the federal government money or, in Washington jargon, reduce 'budget exposure', by substituting CRP hectares for base hectarage on which deficiency payments to farmers are paid. However, base bite turned out to be more of a curse than a blessing because it also increased the opportunity cost to farmers of enrolling land in the Reserve; every hectare put into the CRP was a hectare subtracted from 'base acreage' on which deficiency payments could have been paid (Taft and Runge, 1987). This inevitably meant that farmers would demand more compensation for putting land into the scheme. At the same time, it was discovered that the ARP which operated alongside CRP was mopping up the least productive land on farms before it could be enrolled in the CRP (because it is in the farmer's interest to idle the most marginal hectares first in order to meet his or her cross compliance requirement). Again, the effect was to increase reservation prices by reducing the area of eligible CRP land and raising the bids received for CRP entry. As Taft and Runge (1987, p. 6) put it:

> the result of CRP participation ... is that additional acres of corn land must be idled. If the additional acres to be idled are more productive than those that would have been idled under the ARP alone, the opportunity cost of removing them from production will also be higher and so will CRP bids. The fact that the CRP is a ten year contract makes putting productive acres into it even less attractive, further lowering the prospect that they will be retired at any but a high bid price.

Some of this inflation would have been offset if competitive bidding had occurred because considerable savings were still available from the elimination of windfall gains. But USDA, worried about the uneven farmer response during the first round, anxious to meet its overall enrolment target and keen to maximize the income support and supply control impact of the programme, decided to accept any bid provided it was at or below a predetermined 'maximum acceptable rental rate' (MARR). According to Clark and Johnson (1990), politics seems to have entered into the process of setting bid pool maximums at an early stage. These MARRs were never publicly disclosed to farmers but nevertheless soon became common knowledge. Instead of competing against each other to get land into the Reserve, farmers now colluded to bid at or very near to the MARR set for their bid pools in order to be guaranteed acceptance. They were bidding for what they knew they could get rather than what they were individually willing to accept. Many more bids were accepted and there was a dramatic falling off in low bids (Ervin and Dicks, 1988). As Table 3.4 indicates, by the fifth sign-up, 92% of bids were within $5 of the MARR and 63% were equal to the MARR (Young *et al.*, 1991). The GAO estimated that, as a result, USDA was paying $298 million a year too much for land reserved in the Mountain and Plain States, where payments were often 200–300% higher than average rental values. They calculated that the USDA could have

Table 3.4. Collusive bidding under the CRP, sign-ups 1–5. Source: GAO (1989).

		Percentage of bids			
Sign-up	Number of bids	$0.01–5.00 below	Equal to maximum rental rate	$0.01–5.00 over	Total
1	44,418	6	3	7	16
2	34,435	29	15	8	52
3	45,430	32	41	9	82
4	101,003	41	42	3	86
5	53,107	26	63	3	92

achieved the same level of enrolment by paying an average $25 per acre instead of the $42 per acre it actually offered. Furthermore, since eligibility conditions had been relaxed, the environmental return on these enrolments was much less than it might have been. Dicks (1987) summed up the problem when he calculated that every hectare enrolled into the CRP during the first four sign-ups had been at a net average cost of $14 compared with the erosion and supply control benefits generated.

By now, pressure on USDA to rebalance the CRP was intense and in the run up to the 1990 Farm Bill there was much debate about what the CRP was for and how its performance could be improved. Critics like Daniels (1998, p. 405) had long since come to the conclusion that 'the CRP is closer to being a subsidy programme than a rural planning endeavour'. Others, like Batie (1990) and Zinn (1993) were realistic about the programme's contradictory make-up, aware that it would never have come into existence without farmer and USDA support. They also conceded the need for several different policy goals to be optimized under any future reorganization. Studies such as those by Young and Osborn (1990b) were now more optimistic in predicting significant offsetting savings in budgetary outlays due to the substitution of CRP for cropland over long periods, the base bite effect notwithstanding. Calculations made after the first 11 sign-ups by Johnson et al. (1994) confirmed that over nine Mha of crop 'base' had entered the CRP, saving the USDA over $8.4 billion in deficiency payments. In addition, there were savings due to the indirect effects of reduced agricultural output on market prices. Because the CRP reduces the amount of land in production, it has had a positive impact on farm prices. Higher prices mean lower average deficiency payments on the land which continues to be cropped – saving an estimated $7 billion over the life of the programme. The conclusion was that 'for every dollar spent on the CRP, roughly 50 cents that would otherwise be spent on commodity supports was saved' (Johnson et al., 1994, p. 46). Despite this, growing agri-environmental concern with the wider off-farm costs of soil erosion, pesticide use and wetland loss was making the Department's preoccupation with the on-farm costs of soil erosion increasingly difficult to justify and was to have profound implications for policy design. After a lag of some years, Crosson and Stout's (1983) study, and work by Clark et al. (1985), both showing that the on-site costs of soil erosion may have been exaggerated compared with the impact on water quality, were taken on board by agri-environmentalists and policy analysts, while the newly formed National Wetlands Forum was publishing its manifesto demanding a 'no net loss' policy for wetlands (Conservation Foundation, 1988). Buttel and Swanson (1986) predicted that one result of this shift, and the reassertion of a public interest in agricultural pollution control which it implied, would be a move away from subsidization in favour of state and federal regulation and control. In the meantime, although USDA had agreed in 1988 to modify its enrolment procedures so that land near to streams, lakes or wetlands would be put into the Reserve regardless of its strict erodibility and in 1989 to target cropped wetlands and cropland vulnerable to scour erosion, there was still a sense that environmental targeting of CRP had to be improved, even if this introduced more discrimination into the way the programme was operated and made it less effective (and politically palatable) as a mechanism for farm income support.

In the shadow of 'Gramm-Rudman', the 1987 Congressional law which requires budget cuts whenever a large federal deficit looms, lobbyists had few illusions that the CRP's budget would be increased. Indeed, the discretionary nature of the Conservation Title made it especially vulnerable to budgetary cuts. According to Reimenschneider and Young (1990, p. 88):

> the confluence of events shaping up for the 1990 Farm Bill debate seems likely to place increasing budget pressure on agricultural spending. The perception of an improving farm economy, coupled with the prevailing view that farm programme spending has been inordinately high, promises to make agriculture a target for significant spending cuts as Congress struggles to shrink the budget deficit.
> In many ways, the 1985 Food Security Act may have sown the seeds of its own destruction. It set in motion a series of policy changes that stemmed the tide of deterioration in the farm sector, but at a record high initial government cost. That high initial cost, plus the added pressure from the budget process, now puts agriculture at a unique disadvantage.

In particular, the effects of the 1988 drought and growing export markets meant that policymakers were likely to put a lower premium on supply control. Indeed siren voices like those of Paarlberg (1989) stressed the need to abandon any form of supply control, including CRP, if American farmers were to compete in a post-GATT world (see Chapter 6). Maximizing the environmental return on any future CRP enrolments – 'getting more bangs for the buck' – came to be seen as a *sine qua non* of the programme's long-term continuation.

The Conservation Reserve Revisited

It was a measure of the agri-environmental lobby's growing influence and political clout, then, that the 1990 Food, Agriculture, Conservation and Trade Act, far from retrenching conservation spending, actually consolidated and refined many of the 1985 reforms. The environmental political background to the bill was very different from 1985. A less indulgent and more vigilant coalition of commodity interest groups and a more fractured conservation alliance made for protracted debate early in the legislative process. Many more conservation proposals were published than before and there was some disagreement about the two key provisions put forward by the coalition – to strengthen swampbuster and establish a Wetlands Reserve (Zinn, 1991). It was very soon clear that agri-environmentalists were attempting to shift the burden of policy away from traditional soil conservation in favour of a much broader set of environmental concerns. Writing soon after the Act's (eventually smooth) passage through Congress, Zinn comments:

> If a Rip van Winkle of resource conservation had gone to sleep in the fall of 1985, when the mission of conservation efforts at USDA focused almost exclusively on soil erosion and water problems to enhance productivity, were to wake up today, he would only vaguely recognise the setting. The 1990 (Farm Act's) provisions are mostly about water, trees and other topics more typically associated with broad environmental concerns.

(Zinn, 1993, p. 258)

Under the Act, the Secretary of Agriculture was empowered to enrol another 1 Mha into the CRP and to set up a Wetlands Reserve Program (WRP), into which at least 365,000 ha of cropland would be enrolled and converted to wetland by 2000. An entirely new set of procedures for accepting land into the CRP was established, based on a calculation of the monetized environmental benefits (including, for the first time, wildlife habitat improvements) likely to be generated relative to payment, offers of land from farmers now to be accepted or rejected on the basis of their 'environmental benefit index'. USDA was further encouraged to give priority to locations like the Chesapeake Bay, Long Island Sound and the Great Lakes where water pollution from eroding soil was known to be having greatest impact on water quality and recreation. It was also charged with using the CRP to take cropland out of production in areas overlying aquifers to improve groundwater quality and to aim to enrol any remaining highly erodible land not likely to be covered by conservation compliance. And, in an attempt to deal with the base bite problem, the Act guaranteed to protect a farmer's base hectarage provided it remained in a conservation use after contracts expired. The WRP was designed to dovetail with swampbuster and an array of other restrictions on wetland conversion that had been introduced during the 1980s. Farmers would be offered payments for up to 10 years in return for reconverting cropland back to wetland and agreeing to have a permanent conservation easement written into the title deed to their land. Easements had previously made their appearance under the 1985 Food Security Act, when the Farmers' Homestead Administration was empowered to reschedule farmers' debts in return for the granting of permanent easements on the land concerned. To qualify, farmed land had to be either highly erodible or in some other sense environmentally sensitive (i.e. it was wetland or possessed wildlife value). Easements had long been advocated by economists like Ervin and Dicks (1988), who recognized the public policy advantages of government money buying long-term environmental guarantees. Their operation here would later be compared favourably with time-limited, open-ended CRP contracts.

The 10th, 11th and 12th CRP sign-ups which followed did much to restore confidence in the CRP, bringing in over 1 Mha of the most environmentally vulnerable land in the country. Comparing these last three enrolments with the nine which preceded them, Osborn (1993) notes that only 27% of the post-1990 reservation was from the Great Plains States against 59% before, with 15% of new enrolments from the new 'conservation priority areas'. Generally, there was a strong shift in favour of the Corn Belt, Delta and Lake States and a bias towards preventing erosion caused by water rather than by wind: 75% of the soil conservation achieved after 1990 being due to the prevention of erosion from this source. Significant gains in wildlife habitat and in tree cover were also reported. For many commentators, however, it was far from reassuring to know that less than 5% of the 14.7 Mha finally fixed in the CRP had been screened in the way agri-environmentalists had originally intended. With budget cuts pending, conservationists were haunted by the memory of the great Soil Bank 'plough-out' of the 1950s when 11 Mha of conservation reserve land had been summarily returned to production on expiry of contracts. A survey of CRP contract holders conducted by the Soil and Water Conservation Society (SWCS) in late 1993 (Osborn *et al.*, 1994), suggesting that as many as 63% intended returning their land to crop production as soon as contracts expired, seemed to confirm these fears. Based on results from questionnaires sent to 17,000 farmers or 5% of all CRP contract holders, the survey gave the only consistent national picture of what could happen to CRP land after the first contracts expire in 1995. As Fig. 3.1 shows, the next most likely destination for the land after crops is a return to haying or grazing (23%), followed by retaining the land under trees (4%) or either keeping it under grass or trees. Compared with an

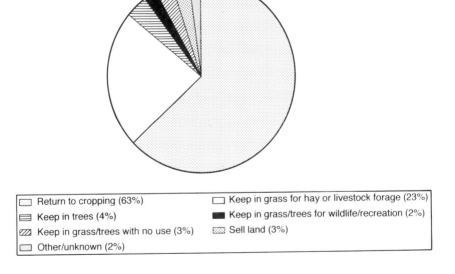

Return to cropping (63%)	Keep in grass for hay or livestock forage (23%)
Keep in trees (4%)	Keep in grass/trees for wildlife/recreation (2%)
Keep in grass/trees with no use (3%)	Sell land (3%)
Other/unknown (2%)	

Fig. 3.1. Post-contract plans for CRP land, 1993. Source: SWCS (1994).

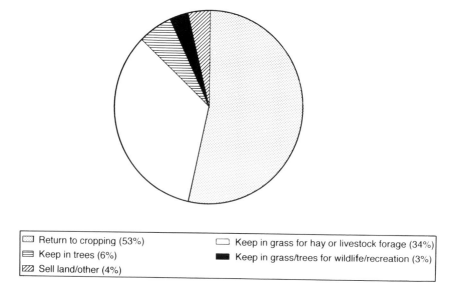

Return to cropping (53%)   Keep in grass for hay or livestock forage (34%)
Keep in trees (6%)   Keep in grass/trees for wildlife/recreation (3%)
Sell land/other (4%)

Fig. 3.2. Post-contract plans for CRP land, 1990. Source: SWCS (1994).

earlier survey conducted along similar lines in 1990 (see Fig. 3.2), 10% more
land appeared to be destined to be returned to cropping. When asked what
would persuade them to keep land in grass, most favoured an extension of the
10-year contracts.

The architects of the 1985 Conservation Title had to some extent planned
for this eventuality. Conservation compliance was supposed to prevent farmers
planting highly erodible land to programme crops without a conservation
plan; Sodbuster took away their commodity support if they did. However,
weaknesses in the implementation of these provisions and deeper flaws in their
operation now began to cast doubt on their likely effectiveness in preventing
an upsurge in soil erosion once CRP contracts expired. To begin with, the conservation compliance standards imposed on farmers by the Soil and Water
Conservation Service (SCS) were probably more permissive than many professional soil conservationists would like. As Clark and Johnson (1990) remark, SCS, used to thinking of itself as the farmer's friend, was anxious from
the beginning to ameliorate the impact of compliance on farmers and, by developing the 'alternative conservation system' standard to achieve 'substantial'
reductions in soil erosion, gave individual officers considerable discretion in
deciding whether or not the standard had been met. Random surveys conducted by USDA suggested widely differing standards and a large number of
cases where farmers were agreeing to comply merely by maintaining a cover
crop on ploughed land during critical times of the year (USDA, 1990). A similar mid-term evaluation of swampbuster revealed that, while the provision

appeared to be arresting wetland conversion, it was not uniformly enforced. The GAO attributed this, with calculated obliqueness, to 'cultural factors' – the difficulties SCS officials were having in adapting to their new quasi-regulatory role (GAO, 1990). Runge (1994) puts it down more straightforwardly to the belief, still widespread among administrators and legislators, that the penalties were too strong. At a local level, where the offending farmer is likely to be a conservation district committee member, it may thus be very difficult to justify 'the death penalty' of withdrawing all subsidies to a farmer who ploughs his erosion-prone grassland. Consequently, while 74% of CRP land could be said to be notionally subject to compliance, it was far from guaranteed that the environmental capital built up over the preceding 10 years would be safeguarded once contracts expired. Meanwhile, the long-term incentive for farmers to comply with even very weak conservation standards was in the process of being undermined by a growth in export markets, rising market prices and declining farmer dependence on commodity support. As Creason and Runge (1990) observe, the problem with compliance is that government agencies find it invidious to operate when it is potentially most effective but ineffective when it is most needed. This is because the sanction of withdrawing commodity support, and hence the imperative to comply, is weakest when buoyant market prices give farmers their greatest incentive to 'plant fence-row to fence-row'. Reichelderfer (1990, p. 220) sums up the policymakers' predicament as follows:

> The consistency aspect of compliance schemes, while laudable, can be their downfall ... sodbuster, swampbuster and conservation compliance were not just made consistent with other farm programs; they were also inextricably linked with them by virtue of the fact that the enormity of the penalty for non-compliance is a function of the attractiveness of the other programs' benefits.

Again, the SWCS survey provides grounds for pessimism. When asked whether they would return CRP land to crops if crop prices were 20% higher than they were in 1993, 78% replied in the affirmative (SWCS, 1994). By the early 1990s, the slow retreat of commodity support in the face of growing demands for a 'decoupling' of US farm policy to give farmers 'the freedom to farm' in a post-GATT world, seemed to many to confirm the wisdom of Brubaker and Castle's (1982, p. 312) remark that 'it may be unwise social strategy to tie long-term concern for soil erosion to an array of agricultural policies that may not deserve continued support on their own merits'.

When Conservation Contracts Expire

Even so, by the time the 1995 Farm Bill came up over the horizon, a less pessimistic but more realistic reassessment had been made of the likely fate of the CRP. With the first CRP contracts having expired and a significant number

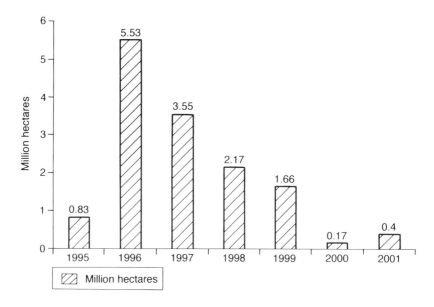

Fig. 3.3. Projected expiry of CRP contracts, 1995–2001. Source: Osborn (1993).

of others due to reach the end of their 10 year term over the following few years (see Fig. 3.3), returning a significant proportion of CRP land to production looked inevitable and might even be welcome if it released money for better safeguarding the most environmentally vulnerable land. Short-term hectarage controls were already being cut back and farmers, looking to exploit the export opportunities of a more open world market, were increasingly anxious to be free of federal government restraint. By conceding that as much as two-thirds of reserved land could be put under the plough without seriously compromising soil conservation standards, the AFT (1996) was acknowledging that a 14.7 Mha CRP was no longer scientifically defensible and could therefore not be sustained politically. Some were even agreeing with Richard Rominger's rather minimalist assessment that the CRP was very much a product of its times, set up in an era of surpluses to ease a farm crisis and deliver some environmental benefits as well.

> We have to recognise that the CRP was deliberately designed by Congress in the 1985 Farm Bill as a transition or bridge to help get the agricultural community through some difficult times ... The CRP provided an almost ideal solution. First, by making substantial direct payments to farmers, Congress was able to use the CRP to support farm income and still reduce target price levels. Second, by retiring so called 'surplus cropland', the CRP would help agriculture reduce unneeded productive capacity ... Without doubt, the major environmental benefits of the program played a large role in convincing Congress this was a

good idea. But those benefits were not the primary purpose of the program. I
believe that if this had been the case Congress would have provided funding for
those activities beyond the ten year contract period.

(Rominger, 1994, p. 54)

Certainly there was agreement that, by failing to provide long-term funding,
the federal government had sent the message that the CRP was supposed
to serve as a transition, though to what precisely was still unclear. As Cook
(1994, p. 64) pointed out, the commodity programmes had been in transition
for 70 years: 'was the conservation transition supposed to end after just ten
years?' Despite some appeals to conservationists to 'defend your base', few
commentators expected that the CRP would survive intact and projections
were made by Osborn (1993; 1994) and others of a steep decline in federal
expenditure on conservation programmes in the years ahead. Annual conservation expenditures had averaged $2 billion in the latter 1980s, the land
retirement payments under CRP and the WRP absorbing some 53% of the
total (Heimlich and Osborn, 1994). A generous assumption, according to
some commentators, was that this subvention would recede to the pre-CRP
average of $1 billion during the 1995–2000 period. Assuming that the CRP
and WRP were allocated 50% of this amount, and that the popular WRP is
maintained at projected levels (having a target enrolment of 395,000 ha by
2000), this would leave $556 million for a slimmed-down CRP.

The question was raised: which land should be retained? It was estimated
that, of the 14.7 Mha in CRP by 1995, about 3.6 Mha or 26% had an erodibility index of less than 8 and could safely be returned to cropping at minimal
environmental cost. Of the remainder, 5.2 Mha were thought to have an erodibility index of between 8 and 15 and thus were likely, under USDA rules, to
be subject to conservation compliance. Most experts believed that such land
should be able to produce crops, provided conservation tillage practices are
adopted and other technical safeguards observed (Lugar, 1994). A further
5.3 Mha was estimated to have an erodibility index of more than 15, of which
21,000 ha were thought to be highly vulnerable to erosion. It was to this core
CRP land that conservation groups like the Conservation Foundation and AFT
now turned their attention, lobbying for a selective recommissioning of CRP
by allowing some farmers to re-enter their land into extended contracts and
for land not previously in CRP to be enrolled for the first time (AFT, 1996).
Heimlich and Osborn (1994) identified up to 8 Mha which satisfied the
requirements of an improved environmental benefit index (see Fig. 3.4), while
the OTA (1995b) argued for an even more tightly defined targeting of any new
land retirements to favour 'environmentally sensitive lands'. Effectively, they
were preparing the case for a smaller, more streamlined programme which, by
being targeted to maximize environmental returns, would leave most
productive land still in an agricultural use. As Taft and Runge (1987) had
proposed some years before, it was time to decouple soil conservation from

Fig. 3.4. Potential future enrolments under CRP. Source: Heimlich and Osborn (1994).

other agricultural policy goals. Meanwhile, conservation opinion seemed to be shifting back in favour of more direct cost sharing of conservation invest-ment on farms and there was growing interest in the scope for using 'complementary technologies' like precision farming and integrated pest management to keep as much land in production as possible at minimum environmental cost (Ervin, 1993; Batie, 1994; OTA, 1995b). Land retirement, if not ruled out of court, was certainly less central to conservation thinking than it once was. As Chapter 6 will show, the agenda was being set for the green recoupling of agricultural support that would mark the next stage in the environmental reform of US farm policy.

For all this revisionism, the 1985 Conservation Title is still seen as one of the most important innovations in the history of US soil conservation policy. In Griffin and Stoll's (1984) terms it was in many respects a 'fundamental decision' which ushered in over a decade of high-level federal government in-tervention in rural America and set in motion waves of incrementalist action that are still having an impact today. As this chapter has shown (Rominger's remarks above notwithstanding), the Title, with its twin provisions for large-scale land retirement and conservation compliance, was essentially environ-mental in its purpose and scope. Conservation compliance was a particularly significant policy departure, imposing on farmers a quasi form of regulation which emphasized the conditional nature of traditional agricultural policy entitlements. In the view of many, this was the most significant feature of the new legislation, challenging New Deal voluntarism and the associated idea that soil conservation was essentially a private responsibility undertaken in order to maintain the productivity of cropland. By agreeing to environment-alists' demands to target the CRP, however, policymakers were also accepting the principle that not all farmers should qualify for conservation subsidies and that conservation priorities, rather than income support or supply control ones, should determine how and where the money was spent. Again, there was an implicit challenge here to traditional thinking which had always justified the wide, and even indiscriminate, allocation of conservation monies on the grounds of the mutuality of farm income support and conservation policy goals.

In the event, a gap between these legislative intentions and policy achieve-ment opened up because of the extensive discretionary powers granted to USDA and its agencies in implementing the provisions of the Conservation Title. As has been shown, this resulted in a dilution of the Title's 'conservation mission' and the assertion of its subsidiary goals of supply control and income support. Significantly, however, the conservation coalition was able to secure a return to a more targeted CRP because, once committed for ostensibly environ-mental purposes, conservation funds had to be shown to be yielding a reas-onable environmental return. They thus exploited budgetary disciplines to promote environmental ones. At the same time, there had been an import-ant coevolution in the nature of public environmental concern, away from the on-site impacts of soil erosion in favour of its off-site effects, which agri-environmentalists were able to exploit in order to bring about a rebalancing of policy and research. This further strengthened the case for targeting and, in a more abstract way, embedded the idea that soil conservation was a societal rather than an individual farmer responsibility. Indeed, the retargeting of the CRP, from a narrow focus on highly erodible land in the early sign-ups to a broader embrace of 'environmentally sensitive lands' in the later ones, illustrates the way the policy adapted to changing public priorities. Whether, in line with the altered view of farmers' environmental responsibilities and liabilities, there will now be a further evolution towards more regulatory solutions to US agri-environmental problems, and if so, of what type, is a question that has yet to be answered.

4 Agricultural Stewardship in the UK

European lobbyists do not enjoy the procedural advantages of regular farm legislation. Opportunities to take stock and introduce omnibus policy packages are rare and there is arguably a greater bureaucratic tendency than in the US for incremental change in the agricultural policy field (Moyer and Josling, 1990). Compared with the US, agri-environmental reform in the EU has certainly been a much more gradualist affair, the evolutionary product of a slower moving policy process. Indeed the UK, which can claim with justification to have been a lead state during the first wave of reform in the late 1980s and the shaper of many of the informing principles behind EU agri-environmental policy, adopted a determinedly pragmatic approach from the start, setting up its first agri-environmental scheme to defuse what was essentially a local conflict between farmers and conservationists and insisting thereafter on the field-trialling and piecemeal introduction of all new agri-environmental policy ideas. As Chapter 2 has shown, the Broads Grazing Marsh Scheme was put together at short notice so that a group of farmers in the Halvergate Marshes – one of the last remaining stretches of grazing marsh in eastern England – could be offered a payment to continue farming in a traditional, environmentally sensitive manner. It was at best an experiment, implemented not by the MAFF but by the Countryside Commission using special powers granted it under the Wildlife and Countryside Act to designate experimental or demonstration projects. At worst, it was merely 'an emergency measure for protecting an area under particular pressure which could not be sufficiently protected through the existing system' (Baldock *et al.*, 1990, p. 144). Before long, however, a reassessment was under way and the scheme was being regarded by conservationists as a prototype for something much more ambitious – the idea of 'paying farmers to produce countryside'. This was rapidly to become a very important idea indeed.

Paying Farmers to Produce Countryside

Countryside management had only recently emerged as a policy goal. While it had long been assumed by British nature conservationists that conservation to maintain biodiversity generally required intervention – Henderson (1992) is one of many foreign observers to be struck by the predilection of British nature conservationists to 'manipulate' nature in order to increase the diversity of habitats rather than to maintain them in their 'natural state' – this was by now practised almost exclusively at a nature reserve level. The natural dynamics which had once operated at a large, landscape scale had been replaced by artificially maintained forms of management on a small scale. In Whitbread and Jenman's (1995, p. 93) words:

> conservation in Britain has relied heavily on traditional approaches to
> habitat management. This is becoming more difficult and expensive
> as management of special sites becomes ever more different from the
> management of the surrounding land. It is also becoming less successful
> in ecological terms.

With the introduction of the BGMS an opportunity had arisen to carry out conservation on a more extensive scale. What had always been implicitly (and erroneously) assumed in agricultural policy circles – that conservation was produced 'jointly' with agricultural outputs – was now to be explicitly engineered by offering farmers contracts to retain low intensity farming practices. Many key features of agri-environmental policy design were prefigured in the BGMS. Its most important innovation was to offer a flat-rate payment to a group of farmers for the purposes of countryside management. Unlike compensation payments available under the Wildlife and Countryside Act, calculated individually on a profits forgone basis and designed to forestall damaging operations, these specified in advance an annual hectarage payment which would be given to any farmer within the designated area who agreed to follow guidelines concerning landscape and habitat management. As Gould Consultants (1985, p. 81) define it, the flat-rate payment is:

> a single payment [which] is applied across an area of land, with no adjustment
> being made for the circumstances of the individual claimant, and with
> no necessary relationship between the value of the payment and the value
> of the individual claimant's losses as a consequence of the conservation
> restriction.

In operation, the scheme was therefore an exercise in conservation on a landscape scale, paying farmers to maintain environmentally sensitive husbandry regimes and farming practices within a tract of countryside which, while still ring-fenced, was large enough to be seen as a distinct farmed landscape as well as a wildlife resource. It piloted the idea that in order to conserve certain valued agricultural landscapes, it is necessary to sustain the farming systems and methods which lie behind them, or to encourage a return to those methods. As the Countryside Commission (CC, 1989, p. 9) put it: 'without the right kind of farming, many of the values of the countryside would be lost'. This put countryside management at the centre of a new justification for agricultural support, establishing the idea that subsidies need to be given to farmers so that they can better fulfil their stewardship obligations. For Selman (1993) it was the first step towards replacing 'a countryside of fragments' with a new and more embracing form of rural estate management rooted in the principles of landscape ecology.

The MAFF, though, had to be persuaded that the scheme would appeal to enough farmers to make a difference before it would commit its own funds to applying the BGMS approach on a broader front. Chapter 2 described the Ministry's initial reluctance to designate other such 'environmentally sensitive areas' on the grounds that this required significant changes in European Community law. By 1985, evidence from the BGMS of a high take-up by farmers persuaded both the NFU and the Minister that the idea could usefully be taken a little further, the Minister making his surprise announcement in the House of Lords and the government mounting its campaign in Brussels for an amendment to the Structures Regulations. Other member states were be-mused at this last-minute intervention by the UK and, according to Baldock and Lowe (1996, p. 14) were more than a little suspicious of the motive behind it:

> partly because it seemed to resurrect predominantly UK Government concerns and partly because of wider suspicions about the British agenda for the future of the CAP. Some Member States were already addressing the kind of problems Britain was encountering either through the creative interpretation of existing European rules or through domestic land use planning controls over the conversion of pasture land. Other Member States saw the British initiative as an attempt to introduce a new subsidy for northern Member States at the very time that southern states were attempting to secure a larger share of the Community market and the CAP budget, and when Britain was pressing for limits on farm expenditure.

Farm ministers, possibly with their minds on other things, nevertheless agreed to the British proposal and wrote into Article 19 of the Regulation a provision allowing member state agriculture departments to offer support to farmers within designated Environmentally Sensitive Areas (ESAs) 'in order to contribute towards the introduction or continued use of agricultural practices compatible with the requirements of conserving the natural habitat and ensuring an adequate income for farmers' (CEC, 1985b, p. 2). This Article, together with the enabling Section 17 of the 1986 Agriculture Act, provided the statutory basis for the ESA programme in the UK.

MAFF now moved quickly, consulting on the criteria by which ESAs would be selected and drawing up the Statutory Instruments which would enable it to offer management agreements to farmers. It was agreed that the designated areas should be:

1. Discrete and coherent areas of national environmental significance.
2. Areas where conservation must depend on the adoption, maintenance or extension of a particular form of farming practice.
3. Areas where the encouragement of traditional farming practices would help prevent damage to the environment.

Clause 18 of the enabling 1986 Agriculture Act further specifies that ESAs should be designated where it appears to ministers that 'the maintenance or adoption of particular agricultural methods is likely to facilitate such conservation, enhancement or protection'. When they were invited to draw up a short list of candidate ESAs, the NCC, CC and English Heritage (EH) commissioned research to identify areas which met these criteria. There were some differences of opinion concerning how tightly the new ESAs should be defined – the NCC believing that boundaries should be widely drawn to embrace a mosaic of habitats and (possibly intensively managed) agricultural land, the CC that they should encompass reasonably homogenous areas of high landscape and wildlife value. In the event, it was the tighter definition which prevailed and the problem of how best to protect areas of scattered or complex environmental interest in the wider countryside was postponed until another day. Despite this, there was broad agreement that the first 'tranche' of ESAs should sample as wide a range of agri-environmental interactions in agricultural landscapes as possible in order to maximize the experimental value of the programme, and that ESAs should be recognizable and regionally distinctive farmed landscapes. It was also agreed that lessons learned from the BGMS should be applied to ESAs. These included the use of a system of tiered payments in which a lower rate of payment would be offered to farmers in order to maintain the broad pattern of land use over the area as a whole and a higher rate to reward more agriculturally restrictive management which safeguarded or enhanced specific features of value. Another was to devolve responsibility for recruitment and advice within each ESA to a Project Officer who would build confidence in the scheme, guide farmers through the recruitment process and monitor compliance.

The Ministry concurred with this analysis and in August 1986 announced the designation of six ESAs in England and Wales, subsequently expanding this to 12 throughout the UK in May 1987. As the Minister declared at the time: 'This new initiative should enable farmers in areas vulnerable to particular pressure to farm in ways which will help them preserve their special beauty and wildlife interest' (Belstead, 1986, quoted in Baldock and Lowe, 1996). Five-year agreements designed to maintain traditional practices, prevent further intensification and, in some cases, encourage modest habitat and landscape restoration, were soon on offer to several thousand farmers occupying over a million hectares of eligible land in what became the biggest single

expansion of the conservation estate in England and Wales since the 1940s
(Potter, 1988). Management prescriptions would be tailored to suit each ESA
and MAFF undertook to publish performance indicators relating to levels of
uptake and the coverage of agreements and changes in the extent of different
wildlife and landscape features. An additional £12 million would be invested
annually in the management of the British countryside in the new ESAs,
which ranged from the South Downs, where agreements would attempt to
reverse the trend of converting grassland to arable and improve management
of semi-natural grassland and valley bottoms, to the Suffolk river valleys,
where they would encourage the better management of wet grassland and
maintain ponds, reedbeds, ditches and dykes, to upland ESAs like the North
Peak, where the objective would be to prevent over-grazing of heather
moorland and conserve landscape features like drystone walls and traditional
barns. In the Cambrian mountains whole farm management agreements
would be offered to farmers to prevent reclamation and afforestation of rough
grazing land, while in the Lleyn peninsula the aim was to maintain existing
rough grazing and unimproved pasture.

   The farmer response was immediate and enthusiastic. Within the first
year of operation, over 100,000 ha had been entered into agreements on 2400
farms in England, representing up to 87% of the total targeted eligible area
(MAFF, 1997; Whitby, 1994). By 1995, after two more rounds of designation
had taken the number of ESAs to 22 and the total ESA area in England to over
1 Mha, 410,000 ha had been enrolled and 4.5% of the utilized agricultural
area of the UK was subject to an ESA agreement of some description (see Fig.
4.1). Compared with the 800,000 ha in total designated as SSSIs (English
Nature, 1996) this was a significant expansion in the area of land notionally
subject to environmental safeguards and eligible for incentives. Farmers have
evidently found ESA agreements to be an attractive proposition and the explanation is not hard to find. Studies such as that by Hill et al. (1992) show
that most participants enjoy financial gains as a result of ESA participation
and that there is a growing awareness of the likely future premium that will
accrue to agricultural land entered into an ESA agreement (indeed, evidence
presented by the Farmers' Union of Wales to a later select committee inquiry
observed that farmers on the fringes of ESAs in Wales were feeling increasingly
aggrieved at the inflation of land values on neighbouring ESA farms). In a
period when livestock farmers especially were caught in the tightening
embrace of a cost–price squeeze, environmental payments became an increasingly attractive proposition, offering guaranteed payments for up to 10 years
without significantly increasing costs (the length of ESA contracts was
increased in 1988). Typically, farmers have an option to enrol land into lower
or upper tier agreements, the latter imposing more environmental restraint in
return for higher rates of payment. Tier 1 payments, for which all ESA farmers
are automatically eligible provided they satisfy the requisite entry conditions,
usually involve farmers agreeing not to intensify production and to conserve

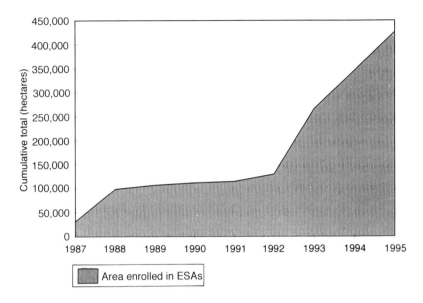

Fig. 4.1. Area enrolled in English ESAs, 1987–1995. Source: Compiled from unpublished MAFF uptake data.

existing landscape and habitat features on their farms. Upper tier agreements, concluded with interested farmers at MAFF's discretion, usually require a definite change in land use and management, for instance, the conversion of arable land to permanent grass. An example would be the tier 1 payments of £130 per hectare, middle tier payments of £215 per hectare and upper tier payments of £415 per hectare available in the Somerset Levels ESA in 1996. According to Whitby (1994), income gains are often very significant, especially when a farmer is being paid to continue with existing extensive farming practices. In the Pennine Dales ESA, for instance, ESA payments have added appreciably to the investment income of farmers (over £230,000 in aggregate during the first 3 years of the scheme) by providing a guaranteed revenue and cutting variable costs. Farmers in the North Peak ESA have enjoyed a £1.2 million boost to farm incomes over 5 years as a result of participation in the scheme, while it is estimated that the net farm incomes of Cambrian mountain farmers have increased on average by £2200 because of the ESA. None of this is to the programme's discredit, of course, since supporting the incomes of vulnerable and economically marginal farmers is in many ESAs essential to prevent farming decline and the loss of countryside management skills. As Whitby goes on to point out, by boosting incomes in a production neutral way, it might be said that the ESA programme has successfully achieved the decoupling of farm income support from production that policymakers are so anxious to achieve. Whether, as MAFF often goes on to claim, the resulting generally high level of farmer participation also means that ESAs have been an environmental success is much less clear.

In the meantime, the fact that more than 87% of ESA land has been entered at the lowest tier (MAFF, 1997), lends support to the view that the programme is only popular with farmers because it does not require them to make any change. Froud (1994, p. 103), for instance, reports that 'participation seems to have had little impact on overall farm strategy ... for many farmers the ESA represents a small proportion of an otherwise fairly intensive arable or mixed farm'. In defence of ESAs, it might be argued that bottom tier entry can build farmer confidence in schemes and, over a period of time, inculcate the more positive attitude towards conservation and land management necessary to make best use of upper tier payments. In the CC's (1997) view, encouraging such a progression is a *raison d'être* of the programme, the aim being to lock farmers into sustainable forms of agricultural stewardship. The NFU (1996) similarly emphasizes the value of 'entry level' schemes such as lower tier ESA agreements which, with their low technical requirements and accessibility, allow farmers to take the first steps towards conservation on their farms. There is some anecdotal evidence that this conservation management progression may have taken place in the Lake District ESA, for instance, though there is little to support the hypothesis that farmers generally are becoming more conservation minded as a result of participation in the scheme (Lobley and Potter, unpublished results). Indeed, McHenry (1996) suggests that there is rising concern within the farming community about the steady bureaucratization of farming as a result of initiatives like the ESA programme. Nevertheless, as Morris and Potter (1995) point out, the extent to which the schemes succeed in bringing about enduring changes in the outlook and attitudes of farmers must be a key measure of their success. Colman *et al.* (1992, p. 69) make much the same point when they argue that 'policy measures which encourage positive attitudes to conservation will in the long term be more effective than those that do not, since a positive shift in attitudes will increase the output of conservation goods at any specified level of budgetary cost'. Indeed, it could be argued that unless they exert such an influence, agri-environmental schemes will be seen as temporary bribes, shallow in operation and transitory in their effects.

In an analytical sense, measuring changes in the attitude and outlook of participating farmers over the period of an agreement has been proposed as an indicator of 'structural learning' by the farmers involved. Work has been conducted (significantly, largely outside the frame of the official socio-economic evaluations commissioned by agriculture departments) which has analysed the motives of farmers entering agreements and sought to compare any differences in outlook and situation with those of non-participants (see, for instance, Morris and Potter, 1995; Wilson, 1997). Taking this more behavioural approach, researchers have been able to get closer to assessing the depth of commitment to, and engagement with, the objectives of schemes, albeit in a static sense, and have made deductions about the likely 'additionality' effects of scheme participation. The starting point for this work was the observation that rates of uptake actually vary widely between ESAs (see Fig. 4.2) and that situational as well as financial factors may be preventing some farmers enrolling in the schemes. For example, while 77% of the eligible area of the Clun ESA was under agreements in 1996, only 17% of the Blackdown Hills and 11% of the Essex Coast ESAs had been enrolled (MAFF, 1997). This variability is of more than academic concern, since a failure to appreciate the importance of non-financial barriers to participation in ESAs where enrolment is low could mean that agriculture departments are missing opportunities to improve uptake by, for instance, improving the promotion or targeting of schemes. (The reported underspend on the ESA budget in 1995/6 was attributed by MAFF to a decline in the rate of applications from farmers to join. But it could just as easily be seen as a self-fulfilling prophecy to the extent that MAFF failed to look beyond financial balance sheets for an explanation of, and solution to, low uptake.) Morris and Potter (1995) were able to place ESA farmers on a 'participation spectrum', ranging from the most resistant non-adopters through a large middle majority of 'compliers' (farmers who conform to the terms of the agreement strictly in order to receive payment) to a small minority of steward-minded farmers who enter the scheme, usually at the highest tiers, in order to extend an established record of conservation investment on their farms. There is little evidence of any movement along this spectrum as a result of participation in the programme, the bulk of current participants apparently joining schemes because of the 'goodness of fit' between scheme prescriptions and their existing farming system. Steward-minded farmers are very definitely in the minority, their participation typically being a continuation of a long trajectory of environmental management on the farms concerned. In 'additionality' terms, policymakers would thus seem to be caught in a double bind. On the one hand, they must set sufficiently undemanding scheme prescriptions to attract enough farmers to make a difference and justify the existence of the scheme in its early years. This appears to have been the MAFF game plan to date. While the result has arguably been a slowing down in the rate of environmentally damaging change on participating farms and the maintenance of existing features and habitat, 'additionality effects' (additions to environmental management and conservation that would not otherwise have taken place) have been hard to prove empirically. There has also been an encouragement of 'compliance behaviour' on the part of farmers (where participating farmers agree to observe the minimal conditions required to qualify for lower tier payments but are resistant to deeper commitments to conservation) which may quickly be abandoned once contracts expire. On the other hand, policymakers now find themselves under growing pressure on value-for-money grounds to bring more farmers into more restrictive (usually upper tier) agreements in order to engineer environmental improvements which go decisively beyond good farming practice. Experience to date suggests that such agreements are

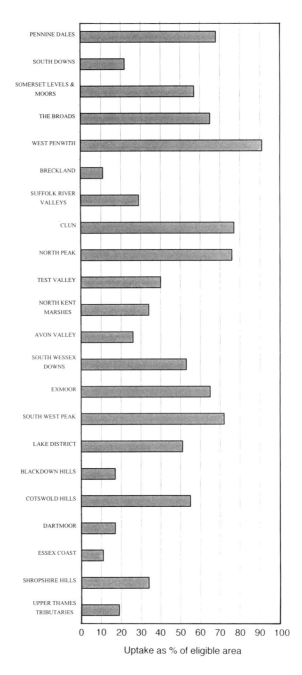

Fig. 4.2. ESA uptake as a proportion of eligible area (management tiers combined). Source: Compiled from unpublished MAFF uptake data.

appealing to a much smaller, self-selecting band of conservationist farmers. While genuinely engaged with the environmental purpose of schemes, it is debatable whether they are making any real additions to the conservation capital of their farm that would not otherwise have occurred.

Assessing the environmental performance of ESAs has not been easy and, more than a decade after the programme's introduction, is still an imperfect art. Countryside management to maintain and improve the ecological and aesthetic value of farmed landscapes is a complex, long-term undertaking which cannot easily be reduced to quantifiable and monetizable outputs. This has not prevented Vail *et al.* (1994, p. 214) predicting that 'the monumental challenge is to classify, quantify and reward the provision of agricultural public goods'. Moreover, the 'audit culture' of the 1990s puts pressure on UK agriculture departments to demonstrate the environmental effectiveness of their agri-environmental schemes and publish information showing environmental value for money. Following the Financial Management Initiative (FMI) of the early 1980s, all UK government spending departments are required to declare their objectives and quantify 'outputs' and even if policy is not quantifiable, must set down their aims and assess their success in meeting them whenever an expenditure of public money is entailed. Significantly, the 1986 Agriculture Act setting up ESAs declares that the minister will pay special attention to 'the effect on the area as a whole of the performance of the agreements' and 'shall publish from time to time such information as he considers appropriate about those effects' (quoted in Whitby and Lowe, 1994, p. 22). The result has been a steady flow of monitoring and assessment reports based on field surveys carried out by the department's Chief Scientist's Group (see MAFF, 1992a–e, 1993a–e, 1996a–e) and the commissioning of independent socio-economic studies to look at the farm business and employment impacts of the schemes (see Russell and Froud, 1991). The former show that ESAs have a rather uneven impact in slowing down or redirecting land use change. Generally, ESAs are seen to have been most successful in maintaining the environmental capital which already exists on farms but have been much less successful in adding to or enhancing that capital. As Richard Wright of English Nature put it: 'ESAs are particularly well adapted to secure and sustain the broader fabric of countryside character over large areas. They have perhaps been less successful in the enhancement and re-creation of the biodiversity resource' (House of Commons, 1997, p. xxiii). In the Pennine Dales, for instance, the ESA is credited with cutting fertilizer use on upland farms but in the North Peak its impact on land use change has been slight. In the Broads ESA, however, agreements have reversed the trend towards converting grassland to arable which was the original reason for setting up the BGMS. ESA payments are also thought to have played a decisive role in halting the conversion of grassland to arable in the Somerset Levels. In the South Downs, agreements have helped maintain chalk grassland and achieved some reconversion of arable back to grass; but they have had a much more tangential impact on the conservation of habitat mosaics in the river valleys there. Even in more successful cases, however, researchers cannot be absolutely confident that they have disentangled the operation of the ESA from other influences which may be pushing self-selecting farmers in the same direction: for example, a reduction in fertilizer use on a farm being run down by an elderly farmer in retirement or a shift out of crops on a farm undergoing restructuring or diversification. It is consequently still too soon to say that ESAs have succeeded in bringing about environmental improvements that go beyond good agricultural practice and thus would not otherwise have taken place on the farms concerned – in the eyes of many, the litmus test of the policy's success as an environmental tool. The tendency of monitoring studies to emphasize the intermediate changes in land use and farming practice that can directly be attributed to schemes – for instance, reductions in fertilizer use, stocking rates or grassland conversion – means that they rarely go beyond measuring the extent to which participating farmers have complied with the prescriptions of schemes. While useful for implementation purposes, this is not a measure of 'additionality'. Meanwhile, there have been reports of a 'halo effect' on some farms where extensification of production on one part of a farm to meet the requirements of an agreement is offset by increased intensity on non-enrolled land elsewhere. An investigation by MAFF revealed that, while there were cases where stock had been moved from agreement land to non-agreement land on the same farm, the scale of movement was small. It continues to keep a watching brief on this phenomenon (MAFF, 1997).

The Promotion of Countryside Stewardship

All this being said, 5 years after the first ESAs were designated, the attitude of policymakers and lobbyists towards the new policy was still remarkably upbeat. At the very least, it seemed as if the ESA programme had drawn the sting of MAFF's critics. The Ministry had introduced what was arguably the most comprehensive system of conservation incentives of any EU member state and, judging by the pronouncements of its ministers (Gummer, 1991), was committed to further improvements and refinements as time went on. This had lowered the temperature of the agri-environmental debate considerably, bureaucratic imperatives taking the place of political ones as the focus of concern shifted away from the broad sweep to the minutiae of policy design and implementation. The late 1980s and early 1990s were a dull time for students of policy change compared with the cut and thrust of earlier debates. In Whitby and Lowe's (1994, p. 20) view 'with the formulation of ESAs, the initiative passed firmly back to MAFF, and conservation pressure groups became marginalised from policy. What many saw as a fundamental attack on the agricultural policy community had thus been deflected'. In one sense however, the support of agri-environmentalists for the ESA programme had always been provisional. It was assumed by groups like CPRE and FoE that the ESA programme was an experiment, albeit a rather elaborate and expensive one, and that the ultimate desideratum was a system which would offer every farmer a conservation incentive of some sort, whether he farmed in an 'environmentally sensitive area' or not (see CPRE, 1990; FoE, 1992). In the words of the NCC Chairman at the time of the first ESA designations, 'we hope that agricultural policy will move in a direction in which conservation is less dependent on the designation of special areas but more broadly based in encouraging conservation management on all farms, wherever they are located' (Wilkinson, 1986).With renewed calls for 'ESAs *sans frontier*' during 1989 and 1990 (see CC, 1989), MAFF officials faced something of a dilemma. Every assessment of the ESA programme to date had shown that any success it might enjoy in terms of high farmer uptake and efficient compliance was due to the way it tailored management prescriptions and payment rates to local conditions. As an early NCC assessment put it 'there is no point in applying exacting prescriptions suitable for excellent examples to poorer areas that will not respond and provide equivalent benefits' (Merricks, 1990, p. 18). It appeared that there was a trade-off to be struck between the coverage of schemes and their environmental effectiveness. There was also a calculation to be made about the political sustainability of ESA schemes, given that measures which cannot demonstrate environmental value for money would inevitably be at risk of being retrenched. How then to deal with the much more scattered conservation resource which was found in the so-called wider countryside and with which conservationists were by this date increasingly concerned?

Resuming its role as policy entrepreneur and risk taker in the agri-environmental policy field, the Countryside Commission (CC, 1989) put forward the idea of a countrywide scheme which would offer farmers and land managers a comprehensive menu of countryside management options varied according to the location and landscape type concerned. Applicants could make their selection from this menu and receive capital and revenue payments in return. The idea would be to emphasize much more the re-creation and restoration of farmland habitats and landscape features than was possible through ESAs and to insist on improved public access as a condition of payment. As the Countryside Commission's influential *Incentives for a New Direction for Farming* paper put it 'environmentally friendly farming needs special incentives in all parts of the country, not just in areas with nationally important landscape types' (CC, 1989, p. 11). The net should therefore be cast very wide. Crucially, however, project officers would be given a power of discretion to decide which offers of land to accept and which to reject in line with their suitability and likely contribution to the broader aims of the scheme. This was entitled 'the discretionary principle' and represented a significant departure from standard MAFF procedure which, in line with EU rules, was always to grant a subsidy to an applicant provided that entry conditions had been met. In justification, the CC argued that discretion would give the administrators of the scheme the ability to be selective in terms of the location and nature of the land accepted but without having to draw rigid lines on a map. Money would thus be more widely distributed than had been possible under the ESA programme, while also ensuring value for money and guaranteeing that policy objectives would be met. More than this, farmers would be encouraged to select from the menu to put together their own proposals for managing any land to be put into the scheme. This reflected a CC view, set out in its paper *Paying for a Beautiful Countryside* (CC, 1993) that the entrepreneurial interest of land managers had to be better recognized in agri-environmental schemes. At a time of growing interest in the application of environmental economics thinking to environmental policy design (Pearce *et al.*, 1989), the Countryside Commission was keen to promote the concept of 'paying for a product', rather than, as with ESAs, subsidizing prescribed farming processes that are deemed to be environmentally sensitive or beneficial.

> We believe that there is advantage to both scheme administrators and scheme
> participants if payment can be linked more to countryside products ...
> (for instance, the conversion of a ploughed field back into flower-rich chalk
> grassland; or the protection of an historic landscape of ridge and furrow, or of
> a heather-rich moorland; or the provision of public access to a coastal hill or
> riverside meadow) ... rather than the management process or practice that is
> supposed to produce them. The public values a thriving and diverse countryside:
> this is much more likely to be produced by schemes that reward land managers
> for clearly defined environmental outputs, rather than for simply following
> a uniform set of management prescriptions.
>
> (CC, 1993, p. 9)

Consequently, under the menu scheme, the CC, acting on behalf of the tax-paying public, would purchase, and farmers, bidding competitively to gain entry into the scheme, would 'provide'. Fraser (1996) argues that the resulting system of 'discretionary bidding' marked a clear change in policy and the first appearance of market-based principles in agri-environmental policy. What is clear in retrospect is that focusing on outputs rather than processes offered a means of achieving a more decisive decoupling of environmental support from agricultural objectives. An important line was about to be crossed which would have evolutionary implications for the nature of farmers' agri-environmental policy entitlements and the relationship between scheme administrators and the recipients of payments.

In 1991, an experimental 'Countryside Stewardship Scheme' (CSS) embodying the discretionary principle, was introduced in England, together with a sister 'Tir Cymen' (TC) scheme for Wales. The objective was to 'sustain the beauty and diversity of landscape; improve and extend wildlife habitat; conserve archaeological sites and historic features; improve opportunities for countryside enjoyment; restore neglected landscape features and create new habitats and landscapes'. Five-year agreements were offered to farmers throughout these countries to these ends. In England priority was given to offers of land from land managers with chalk and limestone grassland, lowland heath, waterside, coastal or upland habitats. The design of agreements exploited many of the lessons that had been learned with ESA implementation and included the use of supplementary payments to reward more ambitious management, and complementary capital grants (absent from ESAs) to fund landscape restoration and access improvements. Stewardship also pioneered the use of advanced market research to set payment rates which would give managers an incentive to minimize countryside management costs and included a form of competitive tendering based on the quality of schemes offered by applicants for a given price. In line with its previously stated intention to develop a more entrepreneurial approach to countryside management, the CC included a provision for defining at the commencement of agreements the 'countryside products' to be achieved and giving special bonus payments at the end of agreements where a new product has been established. Participants were to be encouraged to seek advice and develop the skills needed to become effective countryside managers given the different constraints and opportunities which faced them and would be encouraged to put together their own proposals for enhancing landscape and habitat and improving public access. It was hoped that the exercise of discretion could create a climate of competition between potential applicants, leading to the framing of more cost-effective and more imaginative proposals for funding. TC, operated by the Countryside Council for Wales (CCW) had a slightly different focus, though philosophically it was from the same stable, emphasizing landscape maintenance and restoration but going further than the CSS in the use of whole farm plans.

Despite modest funding (total expenditure on CSS through DOE subventions was less than £52 million between 1991 and 1995, and in 1996, following the assumption of MAFF control, was only £17 million compared with the £49 million expended on ESAs in that year), extremely high take-up rates have been reported in all the target landscapes. A total of 92,500 ha had been enrolled in CSS agreements by the spring of 1996 (see Fig. 4.3). A Land Use Consultants assessment (Land Use Consultants, 1995) had earlier concluded that the scheme had provided good value for money and was ready for further expansion. It argued that the exercise of discretion had been a significant determinant of success and that, as a result, over £8 million worth of potentially low yielding agreements had been screened out of the system. The CC estimated that 12% of remaining calcareous grassland, 10% of salt marshes and 26% of lowland heath had come into the scheme, together with large areas of unimproved lowland wet grassland and traditional orchards. Less clear was CSS's success in field trialling the purchaser–provider model which had been one of the CC's central motives in setting up the original pilot scheme. Research conducted by Lobley and Potter (1997, unpublished results), comparing the ESA and CSS approaches in south-east England, found that, while there was some complementarity between the schemes in terms of the sort of people attracted to join, there was little difference in their *modus operandi*, CSS

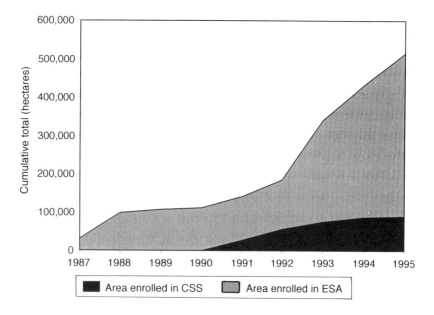

Fig. 4.3. Cumulative enrolment in the English ESA and Countryside Stewardship Schemes. Source: Compiled from unpublished MAFF uptake data.

farmers still tending to enter into prescribed agreements, albeit with greater flexibility as to the options followed. By now, however, CSS had come to be regarded by most agri-environmentalists as potentially the most important of the MAFF schemes. In CPRE's (1997, p. 86) view:

> the CSS has great significance for the future management of the countryside, particularly if the scheme is seen as a basis for a long-term shift in resources away from agricultural production subsidies to direct payments for environmental land management. We believe it provides a useful model for the development of a more comprehensive, nation-wide land management scheme that secures environmental benefits.

Incrementalism Versus Reform?

All this being said, CSS, along with other agri-environmental schemes, still had to be set in the balance against agricultural support under the CAP. Even if seen as pilots for some more ambitious arrangement which could one day replace the existing system of production subsidies (see further discussion below), for the present they were cutting against the grain of an agricultural support system that was overwhelmingly dedicated to encouraging output through further intensification. This at least was the argument now to be put forward by conservation groups like the Royal Society for the Protection of

Birds (RSPB) and CPRE following a decision in the early 1990s to reactivate the larger debate about the environmental reform of the CAP (see Woods *et al.*, 1988; CPRE, 1990). Expenditure on price support and production aids, it was pointed out, continued to outweigh the still very small sums available for agri-environmental schemes. In 1995, for example, the MacSharry compensation schemes cost an estimated 6 billion ECU, agri-environmental expenditure by the EC absorbing just 3% of the total farm budget in that year. Total spending on the CSS was worth less than the subsidy on oilseed rape available to English farmers in that year. This imbalance not only made it more difficult to attract enough farmers into voluntary schemes to make a difference, it also maintained the incentive for intensification on the majority of agricultural land not enrolled in agri-environmental programmes (Jenkins, 1990). Evidence from the Countryside Survey 1990 suggested a continuing loss of biodiversity on arable land and an acceleration in the rate of hedgerow removal and decline during the latter 1980s (Barr *et al.*, 1993). If anything, general standards of countryside management outside ESAs had declined, with a 69% increase in the length of 'non-stock proof' hedges and further deterioration in woodland management being recorded by the survey. The system of Hill Livestock Compensatory Allowances and livestock premiums available to upland farmers was once again at the forefront of environmentalists' attention, groups such as FoE (1992) arguing that if these had been 'greened' in the way conservationists had previously suggested (see Chapter 2), upland ESAs might not have been necessary in the first place. Meanwhile, the introduction in 1988 of a voluntary set-aside scheme narrowly dedicated to production control had underlined for many campaigners the marginal status of agri-environmental schemes and the continued preference of ministers for compartmentalized policymaking. The same Council Regulation which allowed member states with ESAs to apply for 25% EC funding, also set up 'extensification' schemes designed to cut surplus production of cereals, beef, veal and wine by 20% (see Chapter 5). A set-aside scheme was one of the chosen mechanisms. As CPRE, a sharp critic of set aside from its first mooting, put it 'this is a retrograde step ... set aside is unlikely to achieve even the narrow agricultural objectives it is designed for' (House of Lords, 1988, p. 117). This measure promised to take up to 600,000 ha of arable land out of production in the UK (and much more in member states like France), targeting the very areas of countryside ESAs (and, later, Stewardship) could not reach (though in practice, uptake rates have been rather disappointing – see Table 4.1). Yet in its original conception it contained only weak environmental safeguards and few inducements to sound management of the set-aside land (CPRE, 1990). Having decided to engineer what could have been one of the biggest single land use shifts that had taken place in rural Europe for decades, policymakers appeared unconcerned about the likely environmental consequences and unwilling, initially at least, to make significant adjustments to the rules governing where land should be set aside and how it would be managed (Potter *et al.*, 1991).

Table 4.1. Land set aside in the EU under the 5-year voluntary scheme (expressed in thousands of hectares). Source: CEC (1995).

	1988/9	1989/90	1990/1	1991/2	Total 1988–1992
Belgium	0.38	0.12	0.22	0.16	0.880
Denmark	NA	NA	4.59	8.27	12.86
France	14.22	39.70	111.65	68.91	234.48
Germany	167.77	52.20	79.85	179.42[a]	479.24[a]
Greece	NA	0.25	0.20	0.21	0.66
Ireland	1.14	0.48	0.18	1.68	3.48
Italy	93.75	234.97	242.76	150.35	721.83
Luxembourg	0.006	0.03	0.05	0.006	0.092
The Netherlands	2.53	5.91	6.66	0.25	15.35
Spain	25.04	13.85	28.26	36.00	103.15
UK	52.09	48.81	28.59	23.20	152.69
Total	356.926	396.32	503.01	468.456	1,721.712

NA, scheme not applicable.
[a] Includes new Länder.

MAFF, to its credit, committed itself to ensuring that 'the management of set aside land will be beneficial to the community at large' (House of Lords, 1988, p. 19) and encouraged the CC to set up a pilot 'Countryside Premium Scheme' (CPS) in seven English counties which would offer top-up payments to farmers who elected to manage their set-aside land in environmentally productive ways. Under what in the US would be styled a 'green ticket' version of conservation compliance, the CPS had created meadowland open to public access, wooded margins around arable fields, wildlife fallow and more specialized habitats on the 7000 ha entered into the scheme by the time it was wound up in 1992 (Ewins and Roberts, 1992). This represented about 20% of all eligible land in the areas concerned, but with most participants having an average of 40% of their land in the MAFF set-aside scheme. The CC's conclusion was that, while far from ideal, this grafting of environmental conditions on to a commodity scheme was one way of exploiting agricultural policy change for environmental ends and that consideration should be given to 'top slicing' for conservation any further land taken out of production. Two schools of thought on future agri-environmental reform now began to emerge. Borrowing directly from the American experience, the RSPB (Woods *et al.*, 1988; Dixon and Taylor, 1990) suggested that more conservation compliance of this sort was the best way to build environmental safeguards into the operation of the CAP, as well as offering an immediate means of boosting participation in agri-environmental schemes. The problem, as they saw it, was that:

> without some link between guidance on how land is used and the main
> production incentives of price supports and guarantees, the effectiveness of the

new [agri-environmental] measures will be limited to 'amelioration' of the environmental and social impacts of price policy. In effect, Europe has two parallel, but potentially conflicting rural policies.

(Dixon and Taylor, 1990, p. 21)

Advocating what Baldock and Mitchell (1995) characterize as an 'orange ticket' version of compliance, the RSPB envisaged a system whereby farmers would be required to enrol land into one of several alternative agri-environmental schemes in order to continue qualifying for direct production aids. Participation in these otherwise voluntary schemes would effectively become obligatory for anyone who wished to retain eligibility for CAP support and, depending on location and environmental potential, farmers would choose between a redesigned suite of schemes which emphasized extensification, land diversion or the maintenance of traditional farming practices. The idea was to capitalize on what the RSPB expected would be a medium-term trend towards more direct support of farmers through production and income aids under the CAP, as well as the wider use of set aside, to forge a link between the market regimes and agri-environmental policy itself because 'only by making price policy sensitive to the needs of wildlife and the environment can there be true unity of agricultural and environmental goals' (Dixon and Taylor, 1990, p. 5).

By contrast, other commentators were beginning to contemplate a future in which there would be no production support for farmers to receive. In *Future Harvests*, Jenkins (1990) argued that policymakers needed to confront the central issue of the high price regime under the CAP. The analysis on which this paper was based pointed not only to the environmental damage of past high price support and to the free-trade imperatives that were leading to the elimination of price support but also to the dangers of an across-the-board cut in support which entirely abandoned farmers to world markets. Rather than withdrawing support altogether, the government should set up a system of decoupled payments, probably modelled on the CSS, which would pay farmers directly for the environmental goods they produced, graduated according to the environmental interest of the land, the sophistication of the management demanded or the resultant changes in farming practice required. These would eventually constitute the only form of support farmers would be eligible to receive. Unlike conservation compliance, which was an essentially short-term accommodation to the existing system of agricultural support, effectively attaching environmental conditions to subsidies inherently designed to encourage intensification, green recoupling of this sort was likely to be both more efficient and politically defensible.

There is a difficulty in attempting cross compliance when the agricultural policy to which the compliance is coupled conflicts with the environmental aims of the scheme. It is clearly better to 'cross-comply' using payments that are not working against the other elements of policy. It is also considered more advantageous to pay farmers directly for the environmental goods they produce, as this is likely to be administratively easier than attempting to do so indirectly, as well as offering fewer opportunities for fraud and evasion.

(Jenkins, 1990, p. 37)

Moreover, argued Jenkins, this strategy was more in line with the tendency of long-term agricultural policy reform, which was towards the elimination of trade-distorting production subsidies.

Inevitably, perhaps, it was conservation compliance that initially held sway as MAFF, with customary pragmatism, looked for ways of attaching environmental conditions to the new production aid schemes set up under the MacSharry reform package of 1992. EU farm ministers had fully embraced the American concept of cross compliance by requiring arable farmers to set aside up to 15% of their arable area in order to qualify for the new compensatory payments accompanying a cut in price support (for a fuller discussion of the MacSharry reforms and their environmental repercussions, see Chapter 6) and had in addition introduced a number of direct aid schemes for livestock producers. The UK Government agreed to impose the requirement on all recipients of the new Arable Area Payments that they should not damage, destroy or remove landscape features and habitats on set-aside land and, in an effort to carry over one of the features of the CC's CPS, also sought a derogation from EU law to allow additional payments to be offered to producers who agreed to open up public access to any non-rotated set-aside land. Other derogations were agreed allowing farmers to manage set-aside land for goose pasture, as sites for nesting birds, otter havens, and to create sandy and damp lowland grassland. Elsewhere, MAFF conceded the principle that livestock payments to farmers should be withheld or withdrawn where environmental damage has been caused and during 1992 re-opened the debate about 'greening HLCAs' (Hill Livestock Compensatory Allowances), eventually agreeing that these should be withheld where over-grazing is taking place and requiring compliance with a voluntary 'Code of Good Upland Management' as a standard condition of payment. It subsequently extended the over-grazing penalty clause to other direct livestock subsidies. During the UK Presidency of the European Union in 1992, cross compliance was tabled as the sole item of discussion at the Informal Council of Agriculture Ministers held in Cambridge that year (Waters, 1994). Subsequently, the Government was able to obtain EU agreement to the idea that recipients of livestock premium payments should be required to observe ceilings on stocking rates. In 1995 the Government could announce that 'we are committed to looking for ways of extending cross-compliance (*sic*) wherever it is practicable and sensible to do so' (MAFF/DOE, 1995b, p. 7).

By now, however, the limitations of compliance as an agri-environmental strategy were becoming clear and MAFF was in the process of reappraising the long-term potential of its more customized agri-environmental schemes (MAFF/DOE, 1995b). An influential report from the IEEP (Baldock and Mitchell, 1995) had given only lukewarm support to conservation compliance, pointing to the political and legal difficulties the Government would be setting itself in attempting to obtain EU-wide agreement and to the self-discrimination that would occur if it chose to act unilaterally. As the authors of the report pointed out, other member states were likely to be reluctant to impose new conditions on farmers so soon after the MacSharry reforms, yet without EU-wide agreement there would be a danger of introducing serious discrepancies in the conditions imposed on farmers in different parts of the Union, with implications for trade and efficiency. The *ad hoc* manner in which conservation compliance had been sought and applied up to that date appeared to confirm its status as an essentially expedient measure. Moreover, there was the danger that conservation compliance would legitimize and thus perpetuate the production subsidies that the UK Government was anxious to see eliminated as part of a more radical reform of the CAP. By comparison, subsidizing environmental management on farms directly through an expanded agri-environmental programme was much more in step with the decoupling approach. As a House of Lords Inquiry put it, 'positive management of the environment to produce particular public goods should be rewarded' (House of Lords, 1991, p. 47). The CAP Review Group (MAFF, 1995a), set up to advise the Minister of Agriculture on long-term CAP reform, reached a similar conclusion, commenting that a case could now be made for an expanded system of agri-environmental payments 'adequately funded and specifically targeted to deliver environmental benefits through farmers' (MAFF, 1995a, p. 9).

This largely confirmed the approach MAFF had been following since the MacSharry agreement, when it had decided to develop a portfolio of schemes, some permanent, others experimental and provisional, apparently with a view to the more strategic development of a decoupled agri-environmental programme. In its implementation of Regulation 2078, an Accompanying Measure to the 1992 MacSharry market reforms which requires all member states to implement an agri-environmental programme (see Chapter 5 below), the UK Government signalled a commitment to developing a series of decoupled measures which, taken together, 'will tackle environmental priorities in a co-ordinated and mutually reinforcing manner' (MAFF/DOE, 1995b, p. 18). The Ministry's view of the Regulation was that it gave sufficient flexibility to allow member states to pursue programmes which reflected national priorities and the diversity of their own environmental circumstances. Unlike other member states, which now proceeded to set up 'horizontal' schemes covering entire national territories, the UK chose to channel most of the additional resources into the existing suite of 'Environmental Land Management Schemes' (ELMS), as the agri-environmental measures were now to known (see Table 4.2). ESAs were confirmed as the centre-piece of the UK's agri-environmental programme, with the designation of six more ESAs in 1993 and a further six in 1994. Additional funding of £70 million was provided, of which up to £2.5 million would be allocated in 1996/7 to subsidize improved public access

Table 4.2. Expenditure on Environmental Land Management Schemes in the UK (£ million). Source: based on data from House of Commons (1997).

	1992/3	1993/4	1994/5	1995/6	1996/7[a,b]	1997/8[c]	1998/9[c]	1999/2000[c]
ESAs	25.0	31.7	33.0	45.6	47.8	44.1	49.3	53.3
NSAs	–	–	2.0	4.1	4.9	4.7	5.7	6.0
CSS/Tir Cymen	6.0	12.2	16.5	17.4	17.0	22.8	27.5	33.0
Habitat Scheme	–	–	0.8	1.7	2.2	2.5	2.6	2.7
Moorland Scheme	–	–	–	0.4	0.5	0.6	0.8	0.9
Organic Aid Scheme	–	–	0.2	0.5	0.7	1.6	1.8	2.0
Countryside Access Scheme	–	–	0.3	0.3	0.3	0.1	0.3	0.4
Monitoring costs	4.1	5.0	5.3	5.6	3.5	–	–	–
Running costs[d]	11.2	12.3	14.7	14.9	13.2	–	–	–
Total	46.3	61.3	72.8	90.5	90.1	85.6[e]	101.5[e]	114.0[e]

[a] Estimated expenditure.
[b] Excludes Wales.
[c] Projected spending for payments to farmers only. Monitoring and running costs not estimated.
[d] Excludes Scotland.
[e] The sum of individual schemes is less than total projected spending as data for Scotland are not broken-down by individual scheme.

on ESA land under the new 'Countryside Access Scheme'. Agriculture departments also set up a number of more specialized schemes to tackle problems or exploit opportunities not covered by its core ESA programme. The Moorland scheme, for instance, pays farmers directly to remove stock from over-grazed land and to adopt appropriate management regimes and represents a de-coupled way to address the over-grazing problem. Under the Habitat scheme, a version of the dedicated environmental set aside long advocated by commentators and agri-environmentalists (see Potter *et al.*, 1991), farmers agree to take land out of production for 10 or 20 years in order to restore or re-create saltmarsh, water fringe and grassland habitat. This scheme is partly targeted at land previously enrolled in the CPS and, while very modest and experimental in scope, is effectively the decoupled alternative to further conservation compliance within the arable regime. Meanwhile, through the designation of Nitrate Sensitive Areas (NSAs), the UK Government further fulfilled its obligations under the EU Nitrates Directive to reduce or stabilize nitrate levels in public water supplies by paying farmers to make changes in farming practice and land use in vulnerable locations. Nitrate Vulnerable Zones (NVZs) had already been designated under the Directive which requires farmers to follow standards of good agricultural practice without compensation. NSAs covered groundwater sources where more radical departures from good practice were required. Ten pilot NSAs had been designated in 1990 and now, under the

Regulation, a further 22 came into existence in early 1995. Finally, under the Organic Aid Scheme, farmers would be offered transition payments where they converted from conventional to organic production which observed the rules of the UK Register of Organic Food Standards. In 1996 a further piece in the jigsaw was slotted into place when MAFF took over implementation of an expanded CSS from the CC. Together with the ESA programme, this now became a core land management scheme, complementing the more targeted application of ESAs by focusing on landscapes, habitats and other features in the wider countryside.

In the autumn of 1996, a House of Commons Select Committee Inquiry was set up to review the operation of the Government's agri-environmental policy a decade after the designation of the first ESAs (House of Commons, 1997). Its conclusions stand as a useful commentary on the UK's programme at an important moment in its evolution. Few conservationists, agriculturalists or officials would disagree with the verdict of the Commons committee that 'despite nearly 10 years' experience with ESAs, agri-environmental policy is still at an innovation, even experimental stage, and farmers, environmental groups and agencies and the Government themselves are still ascending a learning curve' (House of Commons, 1997, p. xxiv). Here, after all, was an attempt to promote the rather surprising idea that farmers should be contracted to produce landscape and wildlife on their farms and so enhance something called 'countryside character'. To quote Ian Hodge (1996, p. 335), 'this was a new venture whereby government seeks to stimulate the creation of interesting and diverse landscapes for public enjoyment'. In the opinion of supporters, the Government's incrementalist approach, first piloting schemes to judge the farmer response and then introducing, piecemeal, measures designed to address different dimensions of the problem, is an understandable and defensible strategy given the novelty and complexity of the underlying concept. Much has been learned and the feedback of experience in designing, delivering and evaluating schemes continues. This has not prevented critics pointing to gaps in coverage, to the underfunding of the programme as a whole and to its unsatisfactory status as an addition to, rather than a substitute for, conventional agricultural support. These are interconnected problems: the poor coverage of arable land and the wider countryside under the programme is partly due to the slim resources allocated to the Government's only countrywide schemes, the CSS and TC. The reluctance of farmers to enrol arable land where suitable schemes are available has much to do with the generous returns available to arable farmers under the Arable Area Payments Scheme (AAPS) and from the marketplace. As the House of Commons committee again observes, 'large swathes of lowland England have been effectively excluded from agri-environmental policy, and it is these parts of the countryside which are most often seen by most people' (House of Commons, 1997, p. xxiv). In upland areas too, competition continues between the livestock support system and agri-environmental subsidies, and there is mounting evidence of over-grazing outside ESA agreements (Wildlife Trusts, 1996). The much greater enthusiasm of environmental groups than MAFF for conservation compliance stems from their conviction that compliance offers an immediate way to influence the way large numbers of farmers farm by inserting environmental conditions into the operation of producer aid schemes. To this extent the long-standing debate about how wide compared with how deep agri-environmental schemes need to be, remains unresolved. In the short run, there is wide support for an expansion of the CSS in order to better promote the management of the wider countryside.

None the less, the government's willingness to commit more resources at the margin to this and other schemes is likely to be constrained by questions of scheme effectiveness and their publicly perceived value for money. The House of Commons sees this as a double-edged sword:

> Enthusiasts argue that the extent to which agri-environmental measures have had limited success in actually enhancing the environment is due to the various constraints on the policy, principally the financial constraints, and that an expansion and extension of agri-environmental schemes, both in terms of funding and geographical coverage, is necessary to build upon what has been achieved so far. Sceptics, on the other hand, can claim that before further public money is committed to a policy which, in many instances, is paying farmers to undo an intensification of production based on other public subsidies, a more tangible demonstration of the public benefits is required.
>
> (House of Commons, 1997, p. xxiv)

The sceptics need to be taken seriously by those who advocate expansion, for they raise a question about the public legitimacy of a policy which often (and quite legitimately in conservation terms) entails paying some farmers to continue with what they have always been doing. Although the agri-environmental policy experiment was supposed to show that an incentives-based approach was best where environmental protection requires active management by farmers, it is still vulnerable to the charge that environmental improvements have been hard to prove. There are also large gaps in knowledge about farmers' motives for entering schemes and hence the nature of their commitment to long-term management and the true extent of any 'structural learning' that may have taken place. Meanwhile, the UK Government's ability to expand agri-environmental policy in a more substantial, long-term sense, depends increasingly on decisions made at a European and even global (WTO) level. The continuing competition for resources and farmer support between the agri-environmental schemes and a still productivist CAP has been a long-standing source of frustration for UK agri-environmentalists. Resolving this problem, by shifting more resources into green schemes, and accelerating the reform of the CAP, is something that can only be resolved in partnership with other member states.

5 | The Defence of Green Europe

The entry of the EU into agri-environmental policymaking came comparatively late in the day. ESAs, according to one Commission official (Delpeuch, 1994), were widely perceived to be a British invention, speaking to very British concerns with nature conservation and landscape protection and thus unlikely to have wide appeal. The DGVI (Agriculture Directorate) view of Article 19 was of an essentially permissive measure which was at best a side deal to the much more important agreement on the modernization of agricultural structures. This was echoed by the Agriculture Committee of the European Parliament when it asserted, in a report on the Commission's proposed Regulation, that 'it would be wrong ... to over-emphasise the environment when it is in fact our responsibility under these Proposals to look after the interests of agriculture' (Official Journal, 1984, p. 225). For southern member states, still fully committed to a productivist CAP and seriously intent on the further development of their agricultural industries, Article 19 was a particularly puzzling departure, reinforcing the northern bias of rural policy by inventing another mechanism for the transfer of Community funds to farmers in the favoured northern states. Indeed, Yearley *et al.* (1994, p. 15) comment that, for a member state like Spain, 'initiatives aimed at conserving traditional landscapes or preventing pollution (were explicitly viewed) as measures which benefit the core, at the same time as impeding the development of the periphery'. Yet within 7 years there would be agri-environmental programmes operational in every member state, underpinned by significant co-financing from the EC. What began as a concern of a handful of northern member states had become obligatory throughout the EU, bringing over 30 Mha of eligible land into one of the largest programmes of environmental incentives ever established in the EU. As in every other area of EU environmental policymaking, there was to be much horse-trading along the way, policymakers in

Brussels attempting, in this case, to address an agenda for farm income support and supply control as well as the often drastically different environmental priorities of individual member states. The EC, and its Agriculture Directorate (DGVI) in particular, would play an important role in initiating the new policy and orchestrating the national response. At various stages in the process, DGVI became a clearing house for ideas and policy proposals, convening Working Groups to address practical problems of policy design and implementation and brokering agreements between member states concerning the eventual scope and emphasis of the programme. This last involved re-examining, and in some senses, reinforcing, traditional ideas about the nature of farming in a managed countryside. The result would be a rather open-ended definition of what agri-environmental policy was supposed to achieve and a mode of implementation much more decentralized and permissive than anything seen in the US.

An Evolving Agenda

To begin with, however, the suspicions of southern member states seemed confirmed as Germany and The Netherlands both followed the UK's lead in setting up ESAs during the next 5 years, designating over 2.5 Mha as 'environmentally sensitive' in order to be able to offer payments to farmers practising environmentally friendly farming. Following passage of Regulation 1760/87, they were further able to claim a 25% reimbursement from Community funds. By 1990, over 4 Mha had been designated as Denmark, Italy and Ireland followed suit (see Table 5.1). For at least three of these countries, ESAs meant extending schemes and programmes that had already been set up

Table 5.1. Land designated under Article 19 of EU Regulation 797/85 (1990). Source: CEC (1991b).

Country	Area designated (ha)	Eligible area (ha)	Enrolled area (ha)	Number of participants
Denmark	127,970	–	28,060	3459
France	114,620	83,000	36,620	–
Germany	2,560,000	1,223,000	291,646	40,780
Ireland	1140	–	–	–
Italy	944,430	820,740	229,359	6038
Luxembourg	2800	600	610	4
The Netherlands	75,800	27,000	26,815	5013
United Kingdom	740,930	396,570	282,351	4997
Total	4,567,690	2,550,910	895,461	60,291

– Data not available.

or prefigured under national legislation. In The Netherlands, for instance, management agreements for the purposes of landscape protection had been extant since the 1970s (van der Bijl and Ooserveld, 1996). Indeed, the Dutch interpretation of the Less Favoured Areas (LFA) Directive to support farmers on conservation grounds anticipated the ESA approach by some years and was an important point of reference for UK reformers (see Bennett, 1984). The Dutch Government's 'Report on Rural Areas' of 1975 had identified up to 700,000 ha of agricultural land with high landscape value and proceeded to propose a system of management and maintenance agreements, as well as the creation of a network of nature reserves (Slangen, 1992). Although a modest scheme (just 5000 ha were actually under agreements in 1985), the policy piloted the idea of paying farmers for landscape protection and had tested the water in terms of the farmer response. Under the maintenance agreements, for instance, farmers were being paid to conserve scenic features such as pollarded willows and hedge banks – a radical idea for its time. By 1989, 4300 such agreements were operational in 68 regions (Slangen, 1992). Implementation of the LFA Directive was linked to this strategy of landscape protection, areas designated on landscape protection grounds periodically being announced to the Commission as LFA under Article 3(5) of the Directive. In Denmark, public grants for landscape improvement had been in place since the mid-1970s, albeit on a small scale (Primdahl, 1996), while in Germany several Länder had experimented with the idea of offering farmers contracts to protect endangered habitats and species of birds (von Meyer, 1988). In Bavaria, a scheme for the protection of meadowland birds was set up in 1983, eventually giving way to a successor KULAP (Kulturlandschaftsprogramm) scheme which was designed to protect grassland and reduce over-grazing. By the mid-1980s, several other Länder were operating small incentive schemes for countryside protection. In Schleswig-Holstein, for example, farmers could apply for grant aid in order to manage and restore grassland habitat, schemes being introduced to promote the conservation of specific types of habitat on farms (Höll and von Meyer, 1996). Now, with Commission funding, it was possible to extend and elaborate many of these measures, offering farmers within ESAs annual payments linked to environmental contracts (see, for instance, Wilson, 1994). If Article 19 was initially regarded with distrust by these member states, by 1990 it was now on its way to becoming an important element in their policies for farmed landscape protection.

The notable exception was France, for whom the idea of paying farmers to be '*jardinieres de la nature*' grated against the popular view of productive farming as the 'green petrol' of the national economy. As Buller (1992) explains (see quotation on page 43), the philosophy behind ESAs was regarded as alien to French agricultural traditions because it assumed farmers had to be paid explicitly by the state to protect the countryside. The popular assumption, rather as it had been in 1940s Britain, was that the land could safely be left in the farmers' care (and additionally, in the French case, to the Chambers of

Agriculture) without any need for government interference. Indeed, even the most intensive farmer was still assumed to be better placed than anyone else to be the natural creator, user and protector of the rural landscape (Hoggart *et al.*, 1995). Consequently, ESAs had still to be designated in France by the end of the decade and the French Ministry of Agriculture was continuing to make decidedly tepid noises about any movement on Article 19 even as late as 1988. Its eventual conversion to agri-environmental policy, and that of other member states like Italy, came with the realization that the Article 19 approach could be used, and was being encouraged by the Commission to be used, to support the incomes of farmers in a period of farming contraction. Commenting on the EC's early excursions into agri-environmental policy, Fennell (1987, p. 72) observes the careful way in which:

> the Commission presents farming as the preserver of the countryside, the provider of employment and the protector of the rural social fabric. It is not that the Commission has lost sight of the main purpose of farming but rather in the present circumstances perceives that to concentrate on the food producing role would not provide a strong argument for the continued support of agriculture on the present scale.

Article 19, in fact, was fully in line with what might be called the Green Europe model of agricultural development that, on another level, the French were so keen to see upheld. As Chapter 2 has shown, this core policy doctrine, dating back to the Stresa conference, promotes the idea that the survival of the small family farm is essential for sustainable rural communities and well managed landscapes. According to the much quoted formulation given by the EC in its Green Book, 'sufficient numbers of farmers must be kept on the land. There is no other way to preserve the natural environment, traditional landscapes and a model of agriculture based on the family farm as favoured by society generally' (CEC, 1991a, p. 9). For many Europeans, this idea connects with a long-standing debate about the undesirable social and environmental consequences of agricultural decline and land abandonment, which in France goes under the heading of 'desertification'. Delorme (1987) points out that one aspect of this is the threat of agricultural decline to the accepted idea of landscape as a cultivated area shaped by agricultural work. In a country where rural society was once such a dominant presence, the reaction to its imminent demise has been to develop what Judt (1996) calls a 'compensatory myth' of its cultural centrality, a French version of the analogous German attachment to *Heimat*. Less abstractly, von Meyer (1988, p. 72) believes the concept of a 'European Common Garden' has wide support throughout the EU. This embodies the idea of:

> a mosaic of diverse landscapes which, for many centuries, have been mainly shaped by agriculture. In Europe, unlike in America, it is extremely difficult to find untouched natural landscapes ... Europe's landscapes are thus 'cultural' landscapes, an important component of its common cultural heritage.

Article 19, with its provision for maintaining traditional farming practices in order to protect landscapes and habitats, is entirely consistent with this approach. Indeed, to the extent that it could be said to be an evolution of Article 3(5) under the LFA Directive, it is directly concerned with upholding the Green Europe principle that the only way to have a beautiful countryside is to ensure its continued occupation by large numbers of family farms (Potter and Lobley, 1993). As the Commission put it at the time, new forms of public support are needed 'to maintain the social tissue in rural areas, conserve the natural environment and safeguard the landscape created by millennia of farming' (CEC, 1985a, p. ii). More to the point, it also identified the scope for using Article 19 measures 'to solve the income difficulties certain farmers might face and at the same time avoid the production of surpluses' (CEC, 1987, p. 25) and began marketing the income support advantages of the scheme much more directly than before (Clark *et al.*, 1997). After all, the Regulation, in its original wording, required that the adoption of environmentally sensitive farming methods should contribute to farming incomes. The French Ministry of Agriculture seemed to have made the same calculation when, in 1989, it began to identify candidates for ESA status and to set up the machinery for implementing Article 19. As Boisson and Buller (1996) observe, there was little attempt to disguise the principal objective of the policy, which was to supplement the incomes of farmers in LFAs or of those embarking on extensification. Measures were targeted at the protection of sensitive biotopes or 'areas of low farming productivity, where farms are in danger of being abandoned' (Metais, 1993). In the latter case, the new ESAs largely coincided with areas of marginal agricultural production in the mountainous regions of the Alps, the Jura, the Pyrenees and the southern parts of the Massif Central. By 1994 France had 34 ESAs covering almost 10% of its utilized agricultural area and was becoming centrally involved in discussions surrounding the further development of the EU's agri-environmental policy.

The Commission had already signalled how this was to be achieved in its paper *Environment and Agriculture* (CEC, 1988). Previously, at a meeting of environment ministers held in Nyborg in the autumn of 1987, it had been agreed that the environment should be progressively integrated into agriculture and that there was now a case for moving towards the adoption of common principles necessary to achieve this. Interestingly, the Dutch had previously tabled a proposal which would have required an EU code of good agricultural practice to be applied to all farmland, with punitive levies imposed on farmers who failed to observe its requirements. This was summarily rejected by farm ministers on the grounds that a subsidy-based (or rather, 'incentive'-based) approach was more suitable where countryside had to be managed to maintain its environmental values (Clark *et al.*, 1997). Incentives for environmental management on farms and for a qualified application of the polluter-pays principle to agriculture had now been endorsed. Article 19, however, was seen to be too narrowly framed to provide the model for an EU-wide policy. As DGVI rather surprisingly put it: 'specific measures [adopted under Article 19] are often not radical enough to have a major significance for the environment ... The measure needs to be much more widespread if it is to bring about a noticeable improvement in the EU's environmental situation' (DGVI, 1990, quoted in Clark *et al.*, 1997). Driven by what Baldock and Lowe (1996) call its 'integrationist logic', the Brussels bureaucracy was evidently keen to extend the scope of agri-environmental schemes in order to bring all member states on board. The passage into law of the Single European Act (SEA) in the previous year had marked an important advance in the status of the environment in all European policymaking. Up to this point, the Commission had been legally bound to pursue the expansionist (and agriculturally fundamentalist) principles of the Treaty of Rome. Now, charged explicitly with a duty to integrate the environment into the EU's common policies, and given new policymaking powers under various articles of the SEA, the Commission began to recognize the potential for bringing agri-environmental schemes under the umbrella of the CAP. Against a background of impending CAP reform, it was also increasingly interested in the potential for using agri-environmental policy to contribute to production control as well as income support (see for instance, CEC, 1988). The drawback with the British model was that the rather specialized activity of countryside management was never likely to involve enough farmers or displace sufficient land to have the income support or supply control effect policymakers were by now so anxious to achieve. Agricultural pollution as the problem, and extensification as the solution, on the other hand, offered a much more politically appealing way forward.

The problem of animal waste disposal had long been an issue of public concern in countries like The Netherlands, where, as was seen in Chapter 2, the decline of mixed farming and its replacement by large-scale, intensive livestock production had created a huge 'manure surplus' and the attendant problems of nutrient leaching and eutrophication (Reeve, 1993). Some time before agri-environmental policy was being talked about, the Dutch Government had set targets for the control of manures, imposing legally binding restrictions on the timing and rate of applications and setting standards for safe storage on farms (Rude and Frederiksen, 1994). Glasbergen (1992) argues that these standards and their scientific and environmental justification dominated public debate during the mid-1980s in a way landscape conservation never did. According to this author, having swept away early on many of the distinctive features of The Netherlands landscape, agricultural intensification had also destroyed any illusions the Dutch might have had about farming and farmers' stewardship role and helped to nurture a critical attitude towards agriculture which made regulation more politically feasible. During the late 1980s, however, the public health effects of pesticide residues and nitrate in surface and groundwater, thought to be principally due to excessive farm chemical and nitrogen fertilizer use and to changes in agricultural practice, were beginning to attract attention elsewhere. Certainly in France it was the water quality issue more than any other which pushed agri-environmental problems up the political agenda, leading to publication of the Environment Ministry's 'Plan Vert' in 1990 and new legislation covering the control and management of water sources (Bodiguel and Buller, 1989). The initial focus of public concern was localized, linked mainly to the use of inorganic fertilizers and pesticides on arable farms in the Paris basin. Boisson and Buller (1996) chart the slow emergence of the water pollution issue in France and the symbolic challenge this posed to the traditional view of agricultural production as a public good. In 1985 the Nitrate Water Programme had been implemented to raise awareness amongst farmers and to persuade them to modify their farming practices. Meanwhile, impact studies carried out by various departments were raising public awareness of the threat to public health from agricultural pollution. Bodiguel and Buller's (1989, p. 217) assessment of these was that, while their 'publicity is very low key ... the emerging importance of environmental issues in those regions affected by agriculture-related pollution leads one to suppose that a profound and rapid shift in public attitudes will not be long in coming'. In Germany the nitrate problem played what Conrad (1988, p. 203) calls a 'pathfinder role' in propelling agricultural pollution to the front of public debate, where 'for a time it ... tended to overshadow other, perhaps more pressing, environmental problems concerning agricultural policy'.

The justification for this rising concern was the allegation that nitrate in drinking water caused methaemoglobinaemia ('blue baby syndrome') and stomach cancer. The epidemiological evidence for this was rather mixed (see O'Riordan and Bentham, 1993), with little scientific support, at least on the basis of currently available time-series data, for the hypothesis that concentrations of nitrate in drinking water pose a significant risk of stomach cancer. Nevertheless, the hypothetical threat was taken very seriously indeed, providing the EC an opportunity to flex its recently acquired environmental policy muscle. As Reeve (1993, pp. 68–69) observes:

> new opportunities provided by the SEA and the Fourth Environmental Action Plan, which states the Commission's intention to protect water from pollution caused by the spreading of manure and excessive use of fertilisers generally, opened the way for the regulation of one of the most controversial side-effects of agricultural activity, namely the leaching of nitrates from agricultural land into ground and surface waters.

There was also the need to harmonize standards in the emerging single market, given that some member states like The Netherlands were well down the road to regulation (Article 100 of the SEA imposes on the Commission a duty to ensure fair competition). The EC had already decided to adopt a precautionary approach, laying down in its Drinking Water Directive of 1980 a maximum permissible limit of 50 mg of nitrate in every litre of public drinking water and imposing on member states mandatory sampling and reporting procedures. Now, under the Nitrate Directive of 1991, member states were further required to define NVZs within which restrictions on fertilizer use, applications of manure and land use would be applied. States were also required to establish Codes of Good Agricultural Practice, to be implemented on a voluntary basis. These codes must refer to fertilizer application rates, storage of livestock effluent and the establishment of fertilizer records and plans. Taken together, the Directives sent strong signals to member states and created a framework of Community law within which to effect agricultural pollution control; the nitrate problem, it seemed, had become 'a symbol of a mismanaged agricultural economy' (O'Riordan and Bentham, 1993, p. 404) and one policymakers ignored at their peril.

Unconvinced by the scientific evidence, the UK Government nevertheless strained to avoid having to implement either Directive, applying for derogations from the Drinking Water Directive and dragging its heels in setting up NVZs (Hill *et al.*, 1989). Seymour *et al.* (1992) remark that nitrate pollution had long been a bone of contention for the farm lobby and government officials here, both groups questioning the EC standard and its public health justification. The NFU's view of the 50 mg l^{-1} standard was that 'it was set arbitrarily and without any scientific base', while the Regional Water Authorities saw no need to abandon the traditional UK 100 mg l^{-1} standard in favour of what they regarded as a 'scientifically unsound' level (quoted in Seymour *et al.*, 1992). Having been a lead state in developing Article 19, the UK was to be one of the last to fully implement the Nitrates Directive, finally moving to set up ten pilot NSAs covering an area of just 15,000 ha in July 1990. Farmers who agreed to abide by land use or farming restrictions here would be fully compensated under management agreements. A further 23,000 ha was later designated as Nitrate Advisory Areas (NAAs), within which farmers were offered information and advised to modify farming practices where appropriate. Significantly, MAFF's implementation of the Directive reversed the Commission's order of priorities as laid down in the Directive, placing prime emphasis on voluntarily bringing about changes in farming practice rather than restricting the volume of nitrogen fertilizer applied within sensitive areas. The provisional nature of the programme was also heavily stressed by MAFF and its supporters. In this sense, NSAs were to be largely policy experiments designed to test, under field conditions, the effect of measured reductions in fertilizer use and changes in land use on groundwater nitrate concentrations.

By contrast, other member states moved quickly to establish what looked to the farm lobby like a permanent regulatory framework for animal waste and nitrate pollution control. The entire territories of The Netherlands and Denmark were declared NVZs in 1990, despite resistance from the farm lobby, while in Germany enactment of the Fertilizer Application Ordinance ushered in comprehensive restrictions on fertilizer use and the spreading of animal manure (Rude and Frederiksen, 1994). In many countries the Directive was used to consolidate already extant codes of good agricultural practice, technical standards and regulations. This was especially true of The Netherlands, where the 'manure question' continued to dominate debate. Unlike the British Government, the Dutch were anxious to implement the Directive to the letter and while the decision to designate the whole country as an NVZ was resisted by agricultural interests, their opposition was quickly marginalized by a solid alliance of scientists, environmentalists and government officials in the Ministry of Environment and the Ministry of Transport and Public Works (Reeve, 1993). It was agreed that the objective should be to achieve a nutrient balance on all farms by 2000, implementation of the Directive being tied into the National Environmental Policy Plan. Earlier, the government's own 'Nitrate Commission' had recommended that maximum application rates for nitrogen fertilizer should be imposed on all farmers and while the 170 kg ha^{-1} maximum specified under the Directive would require significant further adjustments to farming practice, it was felt to be consistent with already laid national plans. A similarly legalistic, rule-based approach was being developed in Germany, where the Water Management Law had been the subject of protracted negotiation (Bruckmeier and Teherani-Krönner, 1992). Various water protection measures had already been instituted in some Länder, the most significant of which was the Water Penny Regulation in Baden-Württemberg. Meanwhile, under the Water Law and the Liquid Manure Regulation, early, if tentative, moves had been made towards restricting farmers' freedom of action in the interests of public health. The legal framework for water pollution control was also expanding in France, Glasbergen (1992) identifying more elements in common with the Dutch than the permissive British approach. In response to the Directive, new authorizations were drawn up to control intensive livestock farming and the spreading of slurry throughout France (Rude and Frederiksen, 1994; Buller, 1996). Farmers would also be subject to restrictions on farming practice on vulnerable land where new groundwater sources were being developed.

To what extent any of these measures succeeded in applying the 'polluter pays' principle to agriculture is a matter for debate. In theory, the required adjustments to farming practice and investment in storage facilities implied by the adoption of Codes of Good Agricultural Practice entailed farmers bearing some of the costs of pollution control. Dubgaard (1993) estimates that Danish farmers would have had to invest 350 million ECU in storage to meet the terms of the Directive in 1990. In practice, all member states undertake to defray some of the costs to farmers through grant aid schemes of various sorts and, in many cases, to compensate for the income effects of changes to land use and management. A study conducted by the IEEP (Baldock and Bennett, 1991) found, moreover, a principled objection to its application in agriculture among policy officials on grounds of the diffuse nature of agricultural pollution sources and the special status of agriculture as a primary industry. Following an amendment in 1987 to the Water Management Law in Germany, for instance, all farmers were entitled to compensation where their farming practices are restricted in water protection zones. Bruckmeier and Teherani-Krönner (1992, p. 76) reflect that:

> even if lip-service is paid to the polluter pays principle in German
> agri-environmental protection legislation, it appears that implementation
> leans towards burden sharing ... The achievement of ecologically orientated
> programmes has relied crucially on bargaining that has deployed financial
> incentives and compensation schemes rather than legal measures.

Even in The Netherlands, where the government has gone furthest towards prosecuting the principle in its National Plan, and had put forward a proposal for a manure levy to be imposed on any farmer who fails to dispose of their manure surplus in the required manner, farmers within protection zones receive generous compensation from the drinking water supply companies to assist them in storing or transporting wastes. Estimates produced by the Organization for Economic Co-operation and Development (OECD) suggest that at the time the Nitrate Directive was being implemented, Dutch farmers shouldered a relatively small burden of pollution control costs overall, equivalent to just 0.8% of agricultural GDP (OECD, 1995). In the UK, of course, paying farmers to control pollution is positively endorsed as a policy principle, justified under the Directive because of the highly discriminatory nature of the NSAs themselves. MAFF's position could be boiled down to the maxim that farmers who found themselves within an NSA deserved to be compensated because of an accident of geography. In effect, MAFF was willing to pay farmers for not applying fertilizers up to a level that would elsewhere be regarded as 'good agricultural practice'. Significantly, the Commission itself had already conceded that farmers should be compensated under the Directive for reducing nitrogen use and making any necessary changes in farming practice (Reeve, 1993). The polluter pays principle should not apply, it was argued, because the geographically specific nature of the problem would mean penalizing farmers by dint of where they farmed, for example on land overlying vulnerable aquifers. Moreover, excessive fertilizer use, and the conversion of grassland to arable, if this was the cause of the problem, had been actively encouraged by public policy in the past. It would therefore be unjust to demand uncompensated compliance with the new legislation within NVZs. Taking this principle a little further through the planned expansion in agri-environmental policy programmes, there now seemed no reason why farmers should not be paid to 'extensify' production on a much broader front, both to reduce agricultural pollution of various sorts and to cut production.

Extensification was not a new idea. It had been the subject of EC Regulation 1760/87, which was designed to reduce input use, stocking rates and encourage the setting aside of land in order 'to achieve quickly reductions in production which might otherwise be secured only as a result of price pressure over a number of years' (House of Lords, 1988, p. 6). The voluntary 5-year set-aside scheme had been classified as a form of 'extensification' in these terms because, in a technical sense, it entailed farmers producing less from their utilized agricultural areas. Most of the land diverted under this scheme was marginal, but the scheme was nevertheless credited with extending the area of grassland and achieving a reduction in chemical use on participating farms (Hawke *et al.*, 1993). Now though, the Commission was seeking to add a more strictly defined form of extensification to countryside management as an activity which farmers could be paid to undertake under agri-environmental policy schemes, identifying in *Environment and Agriculture* the need for extensification schemes which would protect the environment on a more general scale than was currently possible under Article 19. By arguing that the environmental problems associated with intensive agriculture are 'just as destructive' as those resulting from rural desertification, and, moreover, should qualify farmers for compensation from the public purse, this paper redefined the focus of agri-environmental concern and identified the opportunities for using agri-environmental schemes to reduce output on arable and livestock farms as well as to supplement income. The way was clear for the next round of agri-environmental reform.

The EU's Agri-Environmental Programme

Proposals for an EU-wide AEP were published by the Commission in 1990 (CEC, 1990). These followed closely on the analysis of *Environment and Agriculture* by advocating a two-pronged approach: first, an extended and less geographically restricted use of Article 19 schemes for managing the countryside in accordance with specific prescriptions ('*cahiers des charges*'); second, the use of extensification schemes to reduce the intensity of agricultural production more generally in order to tackle problems of pollution. As the Commission put it:

> the gravity of the situation, in particular as regards water pollution, calls
> for rules to be adopted concerning the use of fertilisers, herbicides or, more
> generally, any chemical practice which may be harmful to the environment.
> [At the same time] it is equally important to adopt a consistent approach to
> safeguarding biotopes and preserving genetic diversity.
>
> (CEC, 1992, p. 66)

This latter principle had by now been enshrined in the EC's Habitat Directive of 1992, with its provision of an EU network of protected sites ('Natura 2000') selected to represent habitats and animal and plant species of special nature conservation interest. Now, under the AEP, subsidies would be available to pay for management and countryside protection on a broader front. The guiding principle was that farmers were now to be regarded as providers of environmental services, for which taxpayers, acting through agriculture departments, were prepared to pay. In the words of the Commission again:

> the farmer fulfils, or at least could and should fulfil, two functions, viz., firstly that of producing and secondly, of protecting the environment ... Concern for the environment means that we should support the farmer also as an environmental manager through use of less intensive techniques and the implementation of environment-friendly measures.
>
> (CEC, 1991a, p. 68)

And as if to underline the growing importance attached to agri-environmental policy, farm ministers abandoned their previous practice of adding an 'environmental' Article to the text of another Regulation when they met in May 1992, agreeing instead to a separate Regulation (2078/92) mandating the creation of an agri-environmental programme. To be accurate, their priority was to agree a package of reforms for the CAP that had been tabled by Agriculture Commissioner Ray MacSharry in early 1992. By comparison with these proposals, the agri-environmental Regulation, along with the other Accompanying Measures relating to forestry and early retirement, encountered little resistance and much support and was agreed before the main negotiations got under way. The Regulation was immediately hailed by John Gummer, UK Agriculture Minister, as 'a major step forward in the integration of environmental protection into the CAP' (MAFF, 1992f, p. 2). According to Baldock (1992), it confirmed the 'stewardship principle' in European environmental policy, a concept now so embedded as to deserve elevation as the fifth principle of European environmental policy.

Under the terms of the Regulation, all member states are required to set up rolling programmes designed to subsidize environmental management on farms. Schemes should apply to all agricultural land, though provision is made for targeting areas of high nature conservation value or acute environmental vulnerability as required. The Regulation sets out the conditions to be attached to the payment of aid under these schemes, referring to the need to take into account 'the undertaking given by the beneficiary and the net loss of income and of the need to provide an incentive'. (Subsequently, Article 9 of Commission Regulation 746/96 would specify that the incentive element should not exceed 20% of the income forgone.) Co-financing by the EC is increased to 75% in Objective 1 regions (defined as areas lagging behind the rest of the EU, where at least 25% of the population have a GDP *per capita* of less than 75% of the EU average) and to 50% elsewhere. Importantly, ministers agreed that the agri-environmental programme should be financed from the guarantee section of the CAP. The Regulation also allows for increases in the 'maximum eligible amounts' (payment rates) that can be offered to farmers under the new schemes. EC money is also available for demonstration projects and for environmental training for the first time. Article 6 states that 'the Community (*sic*) may contribute to demonstration projects concerning farming practices compatible with the requirements on environmental protection, and in particular the application of a code of good farming practice and organic farming practice' (CEC, 1992).

Agriculture departments in member states were now obliged to implement schemes in order to achieve:

1. A substantial reduction in the use of fertilizers and/or plant protection products, or maintaining reductions already made, or the introduction of organic farming methods.
2. A change, by means other than those referred to in (1), to more extensive forms of crop, including forage production, or the maintenance of extensive production methods introduced in the past, or conversion of arable land to extensive grassland.
3. A reduction in the proportion of livestock per forage area (i.e. livestock extensification).
4. The use of other farming practices compatible with the requirements of protection of the environment and natural resources, as well as maintenance of the countryside and landscape, or to rear local breeds in danger of extinction.
5. The upkeep of abandoned farmland or woodland.
6. Set aside for at least 20 years for purposes connected with the environment, in particular the establishment of biotope reserves or natural parks for the protection of hydrological systems.
7. Land management for public access and leisure.

This rather eclectic list reflected a desire to broaden the appeal of agri-environmental policy by identifying extensification (articles 2a–c), countryside management (articles 2d–f) and the prevention of desertification (articles 2g and h) as subsidizable activities. In France, for instance, the ability to use the Regulation to promote extensification and prevent rural depopulation guaranteed it would receive a much warmer reception than Article 19. Boisson and Buller (1996, p. 121) comment that:

> from the outset, Regulation 2078/92 coincided more closely with existing French policy concerns than Article 19 of Regulation 797/85 ever did. Not only is it more relevant to the current preoccupations of a post-CAP and post-GATT agricultural community but it also offers legitimisation to an extensive post-productivist, or indeed a peasant model of agricultural development, particularly in those areas where the productivity drive of the last 30 years has contributed to land abandonment and general agricultural/rural decline. Indeed, as an accompanying measure to CAP reform, the Regulation forms part of a broader rural–agricultural policy, that by encompassing surplus reduction and rural depopulation, falls fairly and squarely within dominant French preoccupations.

In southern member states, meanwhile, the new Accompanying Measures, of which the agri-environmental programme was a part, came to be seen by farming unions and regional officials as offering a supplementary source of income at a time when agricultural activity in marginal areas seemed closer to extinction than ever before (Garrido and Monyano, 1996). Within 3 years all member states would have an agri-environmental programme in preparation or on stream. These comprised:

1. Schemes to reduce nitrate and pesticide pollution so as to protect water supply aquifers.
2. Measures to extensify arable farming by reducing inputs, and livestock farming by reducing stocking densities.
3. Conversion and maintenance payments for organic farming.
4. Schemes to encourage conversion of arable land to grassland, wetland, coastal marsh and river margins.
5. Payments to farmers within ESAs to protect landscapes and wildlife habitats.
6. Measures for maintaining and improving cereal steppelands.
7. Schemes to protect perennial crops of cultural, landscape and wildlife value such as olive groves in Greece, Portugal and Spain, and old orchards in the UK and Germany.

By October 1996 the Commission had provided an estimated 1.4 billion ECU or 3.6% of total farm support expenditure to co-finance these schemes and projected that it would invest a further 4.3 billion ECU over the next 5 years of the programme. Estimates of total expenditure, including national contributions, are not published, but it has been calculated that this latter probably amounts to double the EC contribution (House of Commons, 1997). As Table 5.2 shows, the distribution of the EC contribution is far from even, with Austria, Finland, Germany and France absorbing 82% of the total budget between them. This rather skewed distribution reflects a number of factors, most obviously the more rapid implementation of the Regulation here and the larger utilized agricultural areas of these big member states (Germany and France contain 47% of the EU's total agricultural area). However, it is also a function of the different implementation strategies being pursued by member states – themselves a reflection of national preoccupations, institutional structures and policy traditions. As Table 5.3 suggests, a basic distinction can be drawn between states choosing to channel most of their allocations into targeted 'zonal programmes' (such as the UK, Denmark and Portugal) and those that have opted for 'horizontal' programmes which offer farmers throughout their territories simple schemes in return for relatively low rates of payment (Germany, France and Austria). From the Commission's point of view, stated by Reinhard Priebe, neither strategy is ideal:

> the problem with the first [being] that these undertakings might be very close
> to what is normal agricultural practice, so that aids might be very close to
> income aids. In the other case you have the problem that you are not easily on

Table 5.2. EU expenditure on Regulation 2078 in 1996. Source: CEC (1996c).

Country	Expenditure on Reg. 2078 schemes (million ECU)	% of EU budget
Austria	541.0	38.9
Belgium	1.5	0.1
Denmark	5.8	0.4
Finland	256.6	18.4
France	118.9	8.5
Germany	231.7	16.7
Greece	1.5	0.1
Ireland	43.4	3.1
Italy	41.5	3.0
The Netherlands	7.6	0.5
Portugal	40.0	2.9
Spain	32.8	2.4
Sweden	43.4	3.1
United Kingdom	25.5	1.8
Total	1391.2	100.0

Table 5.3. Area designated under EU Regulation 2078 (1995). Source: CEC (1996c).

Country	Area approved for EU funding (thousands of hectares)	Area approved for funding as % of country UAA	Area approved for funding as % of EU designated area
Denmark	44.2	2.1	0.4
France	5061.6	16.8	42.5
Germany	5380.0	31.3	45.2
Ireland	56.2	1.3	0.5
The Netherlands	16.1	0.8	0.1
Portugal	471.3	14.8	4.0
Spain	89.8	0.4	0.8
United Kingdom	796.2	4.6	6.7
Total	11,915.4	11.8	100.0

the line of the Regulation, which says that the schemes have to be applied all over the member state everywhere.

(House of Commons, 1997, p. xv)

Indeed, having expected each member state to draw up a single national scheme, possibly modulated in order to address problems specific to location, where appropriate, the Commission quickly found itself confronted with 117 different schemes, some zonal, others horizontal. In Germany, France and Austria, a very high proportion of the utilized agricultural area is eligible for agri-environmental support, various 'entry level' schemes having been drawn up for the purposes of maintaining existing low intensity farming or to bring about an extensification of production, invariably by paying farmers to reduce input use or stocking densities. Under the French *'prime à l'herbe'* (grass-land premium) scheme, which applies to the whole national territory, live-stock farmers are required to agree not to exceed a stocking rate of more than 1.4 livestock unit per hectare and to maintain the existing area of perman-ent grassland on the holding. Landscape features such as hedges, woodland and water-side habitat must also be managed and maintained. Enrolments into this scheme currently account for 83% of all land entered into agri-environmental schemes in France (Buller, personal communication), with key concentrations in the extensively farmed Massif Central (see Fig. 5.1). In Germany, the Gemeinschaftsaufgabe Verbersserung der Agrastruktur und des Küstenschutzes (GAK) acts as a federal umbrella for a variety of extensifica-tion, grassland maintenance and organic farming schemes implemented at Länder level. Again, the aim is to bring large numbers of eligible farmers into incentive schemes. As Höll and von Meyer (1996, p. 83) put it: 'German implementation is not restricted to specific ESAs, as in the case of Britain ... [rather] the entire German territory is considered environmentally sensitive, although to differing degrees and for different purposes'. Zonal programmes have been set up in both of these countries – in France there are various water protection measures and schemes aimed at the conservation of particular agricultural landscapes and the preservation of rare breeds, while in Germany the Länder have drawn up targeted schemes for grassland management and habitat protection – but these pale in budgetary terms beside the horizontal programmes.

By comparison, nearly all of the UK's and Denmark's allocations are being funnelled into targeted ESA-type programmes. As Chapter 4 described, the approach of the UK's MAFF has been to continue the agri-environmental policy experiment, setting up a raft of schemes in order to maximize the opportunities for learning but also investing in the strong by putting most money into the flagship ESA programme. In common with Denmark, there is a strong belief that schemes need to be targeted in order to maximize environmental value for money. The Danes have continued to expand their ESAs (more fragmented and smaller than the UK's) and to place much emphasis on grassland conservation and reconversion (Primdahl, 1996). Ireland's Rural Environment Protection Scheme (REPS), though similarly habitat and landscape orientated, applies to the whole country and has been criticized by Irish environmental NGOs, who lobbied for a zonal programme (BirdLife International, 1996). The other member state closest to the UK and Denmark in giving priority to landscape protection is Sweden, which has used Regulation 2078 to set up a compre-hensive system of measures to conserve traditional meadowland habitat and to

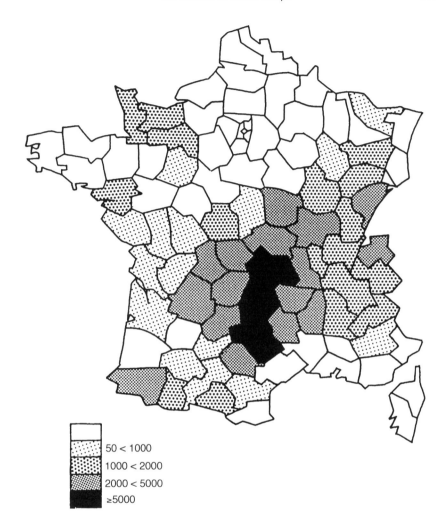

Fig. 5.1. Number of agreements in French grassland premium scheme, 1992.
Source: Buller, personal communication.

maintain the open agricultural landscapes characteristic of northern Sweden (Rundqvist, 1996). Unlike the UK, however, eligibility is generously defined, the Landscape Conservation Measures, Sweden's equivalent of the Country-side Stewardship Scheme, being one of largest programmes of landscape conservation funded under the Regulation. Up to 50% of Sweden's utilized agricultural area is thought to be eligible and farmer participation rates are among the highest of any 2078 programme. Austria and Finland, the other two new members of the EU, have largely followed the extensification route and both are major recipients of EC agri-environmental programme funds.

Austria's ÖPUL ('national environment') scheme shares many of the characteristics of the French grassland premium scheme, covering 92% of the country's agricultural area. Farmers who enter the basic scheme are again required to observe stocking rate limits and to maintain remaining pastureland and landscape features on their farms. The overriding objective of the Finnish General Agri-Environmental Protection Scheme (GAEPS) is 'to reduce pressures on the environment, especially on surface waters, groundwater and air and to reduce hazards caused by the use of pesticides'. Southern member states, especially Greece and Portugal, have been among the last to implement the Regulation, largely for institutional reasons. The Greeks had obtained approval for just one programme by the end of 1996, with a horizontal scheme in the pipeline for the promotion of organic farming and a zonal one targeted at reducing nitrogen pollution in the Thessalia Plain. For Portugal and Spain, agri-environmental policy does not sit easily with the productivist thrust of agricultural policy since accession and there have been delays here too in setting up the national and regional structures required for implementation. Not surprisingly, measures to support extensive systems of arable and livestock production absorb the lion's share of money currently being invested in these southern states. In Spain, however, significant sums have also been allocated to zonal programmes aimed at aquifer protection and water conservation. The Portuguese are implementing, or have proposed, schemes on a regional basis. The only regional programme approved to date, the Castro Verde Programme, covers just 1% of the national territory but is focused on an area of very high conservation importance.

Dark Green or Pale?

Given its provenance, it would be naïve to regard the AEP as purely environmental in either its conception, design or implementation. In agreeing that Regulation 2078 should be an 'Accompanying Measure' to the MacSharry package of CAP reforms, farm ministers were as much influenced by the income concerns of agriculturalists as the priorities of agri-environmentalists. As a Commission official was at pains to emphasize:

> it is very important to see this measure in the context of the overall reform of the CAP of 1992 ... It is not said in the Regulation that it is exclusively an environmental instrument. It is certainly not exclusively an income aid for farmers. It is something which has several objectives, and I think this is very important to keep in mind when you talk about this scheme and when you apply it.
>
> (quoted in House of Commons, 1997, p. xv)

This is borne out in the status of the Regulation as a 'common measure', whereby co-financing is permitted but only on the grounds that schemes have 'a direct link with the improvement of [agricultural structures], the ration-alization of farming practices or the ensuring of a fair standard of living for the agricultural population'. Of course, according to a Green Europe view, there should be no necessary conflict between these objectives, using the Regulation to maintain the incomes of marginal producers in order to prevent land abandonment, desertification or landscape decline, being a perfectly defensible use of public funds. Hence this positive endorsement by a key Commission official of the Regulation's multiple objectives. Others are less convinced by this clever conflation of policy objectives, pointing to the marginal status of the Regulation in the overall MacSharry package and questioning the advisability of intermixing income support and environmental protection policy goals. For some, Conrad's (1990b, p. 18) assessment that 'nowhere has the core of estab-lished agricultural policy yet been touched by environmental policy measures' continues to be true, even in the wake of Regulation 2078. These critics point to the fact that, despite the injection of new funds, agri-environmental spend-ing is still massively outweighed by continuing price and production support under the CAP. There is thus an inevitable tension, if not outright contradic-tion, between the large direct and indirect payments being channelled to farmers under the MacSharry reforms and the small conservation dividend trickling through 2078 (expenditure on price support and compensation pay-ments totalled 39 billion ECU in 1996 compared with the 1.4 billion invested in agri-environmental programmes). On the other hand, any expansion of agri-environmental policy, as it is currently constituted, risks the further cor-ruption of environmental schemes by agricultural interests as policymakers seek to exploit ambiguities in the wording of the Regulation itself to develop new forms of income support. Wilkinson *et al.* (1994, p. 22) are particularly exercised by the decentralized, permissive implementation of the programme and ask:

> has the environment become a lifeline for an embattled agricultural policy, allowing each Member State to support its farmers according to national objectives? Many, including everyday, farming practices can be considered to have environmental benefits. With Regulation 2078/92, Community (*sic*) provisions now allow Member States very considerable freedom to set their own environmental objectives, and, to a degree, income objectives, which can then be implemented with Community co-finance or as purely State-aided measures.

Tangermann (1992, p. 20) puts it more strongly, detecting 'a strong tendency to argue for all sorts of measures which are called environmental policies, but whose objective essentially is to find new ways of channelling money to agriculture'.

Actual evidence for this subtle subversion of agri-environmental policy is not easy to find. It is true that member state agriculture departments enjoy considerable latitude in deciding how to set up and implement the 'multi-annual zonal programmes' required by the Regulation and that there is no requirement for it to be applied in a common fashion. This is intentional, the Commission having decided that member states should be given maximum scope for tailoring policies to local conditions according to the principle of subsidarity. In Baldock and Beaufoy's (1992a, p. 2) view, the EC:

> has been brave enough to suggest that most of the agri-environmental measures should be developed and implemented at a regional level and that a detailed Community framework is not required. This is a commendable response to criticisms of earlier policies and provides the opportunity for developing policies which can be adapted in the light of local experience.

On the other hand, a price appears to have been paid in terms of a loss of central administrative control and there is justifiable concern that many of the new measures, particularly those 'horizontal' schemes promoting extensification, are designed to be more efficient in supporting the incomes of farmers than protecting the environment. Looking at the implementation strategies of some member states, it is clear that a game is being played which has as one of its objectives the maximization of EC receipts. The failure of most countries to define performance indicators is a growing cause for concern in official circles, as is the wide variation in scheme prescriptions and eligibility criteria. Relatively small investments are being made in the monitoring and assessment of schemes, raising further questions about the likely accuracy of any assessments that are undertaken and of the likelihood that schemes will be refined over time. Above all, it is far from clear that all member states have a clear environmental rationale for the programmes being set up, the lack of reference to the Natura 2000 network in the targeting of habitat protection schemes in some countries being indicative of this (BirdLife International, 1996).

So far as extensification is concerned, the problem begins with the wording of Article 2 of the Regulation, which talks about a 'substantial' reduction in input use. What constitutes a substantial reduction is not defined, though the implication is that it will go beyond what might be considered good agricultural practice: for instance, the elimination of wasteful, 'insurance' applications of fertilizers and farm sprays. The environmental justification for cutting back on fertilizer use has never been very clear, particularly where the extensification is taking place across the board and involves lowland arable farmers in northern Europe reducing fertilizer use from, say, 250 kg ha^{-1} to 200 kg ha^{-1} year^{-1} (Baldock and Beaufoy, 1992a). As these authors put it 'there is little purpose in expending part of the limited EC AEP budget in order to promote a general and unspecified reduction in input use' (Baldock and Beaufoy, 1992a, p. 7). Moreover, there is the danger that even after 'extensification' has taken place, some arable farmers will still be applying inputs at rates that are still very high in absolute terms. Delpeuch (1994, p. 41) compares the situation to that of the reckless driver who, even if he reduces his speed from 120 to 100 miles an hour, 'would still not deserve any recognition from society – why should we pay farmers who do much the same thing?'. Leaving to one side the subversion of the polluter-pays principle being implied here, policymakers may face presentational problems in justifying schemes which pay farmers who have already intensified with the aid of CAP subsidies to reverse the process. Putting this more strongly, it might be objected that subsidies to reduce input use will actually tend to provide more opportunities for those farmers who have gone further along the intensification route. 'Why should farmers be paid to undo what they were previously paid handsomely to do?' asked the Environment Committee of the UK House of Commons (House of Commons, 1997). There is no proof that this is happening, but in Germany the GAK permits the setting up of schemes which appear to be paying farmers to make reductions in fertilizer and farm chemical use irrespective of their starting point or of the likely environmental impact of any changes made (de Putter, 1995). The position is similar in Finland, where the GAEPS puts heavy emphasis on reducing fertilizer and chemical use in absolute rather than relative terms.

Regarding livestock extensification, the Regulation refers to farmers who 'reduce the proportion of sheep and cattle per forage hectare'. Again, there is a theoretical possibility that farmers who merely reduce stocking rates from a very high level to a slightly less excessive level will qualify for payment. In practice, farmers who stock above six livestock units per hectare are disqualified under the Regulation and the final stocking rate achieved must be sustainable in environmental terms. More problematic here, and under other provisions of the Regulation, is the scope for paying farmers to simply maintain existing stocking rates and farming practices. The original environmental justification for these maintenance payments is clear: to keep in place low intensity farming systems essential for the conservation of farmed landscapes and farmland habitats. Unfortunately, there are signs that the rather permissive wording of the Regulation has been exploited by some member states to offer blanket subsidies to livestock farmers *per se*, regardless of the environmental value of their farms or the extent of any changes in farming practice and land management achieved. The French Ministry of Agriculture's *'prime à l'herbe'* scheme and the Austrian's grassland extensification scheme are possibly the most pronounced cases of this, each introduced for entangled agricultural and environmental reasons. The French scheme was partly invented to compensate livestock farmers for the very high compensation payments available for maize silage under the MacSharry arable regime and to buy their consent for the larger package of livestock reforms (Baldock and Lowe, 1996). The Austrian Ministry of Agriculture argued that its extensification scheme was necessary to prevent livestock farmers intensifying production following Austria's accession to the EU and the sudden availability of generous producer aids.

There are two sides to this argument for supporting farmers as a precaution against further intensification. On the one hand, applied as a general principle, it increases the susceptibility of the policy to 'moral hazard' whereby support is given to farmers who have no intention, or are under no pressure, to intensify production. If additionality cannot be proved, the policy may eventually lose public support and be withdrawn. On the other hand, a precautionary use of agri-environmental policy schemes needs to be generally upheld because of the importance attached to keeping high natural value farming systems in place. Agri-environmental policy will have served a useful purpose if, as McCracken and Bignal (1995) suggest, it underwrites traditional farming systems, especially in southern member states. But to do this requires measures which effectively operate as income supports, regardless of what economists call 'adverse selection' in the short term.

This is a familiar conundrum, but one which is particularly problematic in the European context. How feasible and desirable is it to maintain a separation between supporting farmers' incomes and paying for the production of environmental goods when the environment is perceived to be jointly produced with the activity of farming? Given the European conceptualization of agri-environmental problems, it seems unlikely that the EU's agri-environmental policy will ever be entirely decoupled from agricultural policy concerns in the way Wilkinson and others appear to have in mind. Policy evolutionists might anyway argue that these issues should be viewed in the context of a policy that is still at an early stage in its development. It was always the Commission's aim to apply agri-environmental schemes as widely as possible and by maximizing the 'reach' of policy, the current extensification measures allow policymakers to exert an influence on the land management practices of large numbers of farms. There are already encouraging signs that Brussels is anxious to improve the 'additionality' effects of AEP programmes, issuing an amending Regulation in 1996 which requires agriculture departments to ensure 'environmental impacts are significant and go beyond what could be considered as good agricultural practice' (Scheele, 1996, p. 4). Meanwhile, there is enormous scope for pooling experience and identifying best practice from the diverse experiences of member states. European networks of knowledge regarding agri-environmental policy are developing fast and there are a number of comparative research programmes in progress which will inform the Commission's review of 2078 that is due to be completed in 1998.

According to Weale (1992), environmental policy is as often about deciding the nature of a problem as deciding between competing interests involved in a problem. In the case of the EU's agri-environmental policy, there has been a particularly clear evolution in thinking about the environmental impact of modern agriculture which has paralleled the development of the policy itself. By initially choosing to focus on the problem of countryside management and landscape protection, policymakers were able to justify giving farmers environmental payments as a reward for the private production of public goods. One of the assumptions being made here was that agriculture departments had a role in supporting traditional farming systems in order to prevent agricultural decline and desertification. This connected with one of the most influential ideas behind European rural policy, that of the desirability of maintaining a Green Europe of small family farms, in which agricultural prosperity is a pre-condition for sustainable rural communities and well managed landscapes. As Eric Fottorino, agricultural affairs correspondent with *Le Monde* during this period, put it 'This new approach to agriculture as a source of social harmony and well being as well as of foodstuffs marked the coming of age of green Europe, where emphasis was placed on farming not as a mere means of production but as a way of life that had been sorely undervalued' (Fottorino, 1990, p. 12). While Article 19 was chiefly a British invention, it was eventually being more widely adopted as member state agriculture departments and the farm lobby began to appreciate its more prosaic but politically useful value as a mech-anism for farm income support. Then, when public concern about agricultural pollution demanded recognition of what economists call a 'negative extern-ality' dimension to the problem, the adoption of a precautionary approach expressed in the concept of 'extensification' allowed national policymakers to continue advocating subsidization in preference to regulation and control. This strategy of paying farmers to reduce their use of organic and inorganic fertilizers and farm chemicals had the additional advantage of offering a means of reducing agricultural output at a time when the control of surpluses was the main driving force behind agricultural policy reform. Subsidized ex-tensification was a Green Europe compatible solution to the problem of over-production because it was supposed to achieve more quickly a cut in production that could otherwise only be achieved through steep reductions in price support and a drastic restructuring of the farming industry.

When the European Commission came to draw up an EU-wide agri-environmental programme to accompany the 1992 MacSharry CAP reforms, the potential for using both countryside management and extensification schemes to support the incomes of farmers was a material consideration, but one that could be squared with the interests of environmentalists, given the model of farmers as providers of environmental services that had by then been widely adopted. Again, farming interests were well served by an evolution in thinking which justified an even wider deployment of environmental payments. It remains to be seen how defensible the resulting, rather loosely constructed, agri-environmental policy will prove to be. Despite the finished appearance of many of the larger national programmes, they are still essentially experiments which have yet to establish their green credentials and environmental value for money. Meanwhile, the EU's trading partners are beginning to ask ques-tions about what the policy is for and how far it can be said to satisfy the criteria that have been drawn up by the WTO concerning such matters. The internationalization of agri-environmental policy is the final chapter in this story.

6 Agricultural Liberalization and the Double Dividend

Agri-environmental reform until now has arisen out of, and reflected back, largely domestic concerns. The budgetary crises of the mid-1980s were absolutely critical in changing the dynamic of the debate about agricultural policy reform. Without this spark, it is unlikely that the greening of farm policy would have ignited when it did (though agri-environmental policies would have emerged eventually). The effect, arguably, was to make national policy-makers much more receptive to the dispositions of environmentalists than they would otherwise have been and, in the search for new ways to justify taxpayer support for the farming industry, to implement some of their ideas by transferring part of the farm budget into environmental schemes. Agriculturalists and environmentalists were also increasingly able to find common cause as the decade wore on. Far from challenging their traditional policy entitlements, arguments in favour of an expanded system of green payments offered the farm lobby a means of defence, provided agri-environmental reform could be presented as requiring a redirection rather than a net withdrawal of farm support. For conservationists, the new measures represented a significant additional public investment in soil conservation, pollution control and countryside protection, and promised a more fundamental redirection of government spending at a later date. As a process, agri-environmental reform looks set to continue, and even accelerate, in the years ahead. Admittedly the budgetary impetus is presently much weaker than it once was, especially in the EU, where the MacSharry reforms of the early 1990s have controlled the trend growth of CAP expenditure. Even Moyer and Josling (1990, p. 210), firm believers in the catalytic role of budgetary crises in agricultural policy change, admit that 'there is a limit ... to the efficiency impetus created by a budgetary crisis. As soon as enough resources have been saved in order to deal with the crisis, pressure is removed to take further action'. Today, though, it is increasingly difficult to find even the most hard-bitten commentator willing to deny that farm subsidies will take on a greener tinge. Rather, a new set of pressures for international agricultural policy reform have emerged following the successful conclusion of the Uruguay GATT Round in 1993. This event, and the momentum for further reform it generates, provides the setting in which future agri-environmental reforms will have to be debated and worked out, as agri-environmental policy is swept into a larger, international debate about the liberalization of agricultural trade.

The internationalization of agricultural policy reform dates from 1987, when, for the first time, OECD governments committed themselves in a Ministerial Declaration to a concerted reduction in agricultural support and the inclusion of agriculture in the Uruguay GATT Round. This was a watershed event, for until then agriculture had always been granted exceptional status and enjoyed generous immunities under GATT rules. Having agreed that this should end, members had cleared the way for coordinated reform. And, indeed, the signing of the URAA in 1993 put in place a common strategy for improved market access through tarification, the gradual replacement of price support with 'decoupled' income payments and the phased elimination of export subsidies. The Agreement means that the agricultural policies of industrial countries are on converging paths, subject to the same internationally agreed rules and procedures. Agri-environmental policy will be no exception to this. The URAA, by placing agri-environmental subsidies in the 'green box', and thus granting them immunity from existing and likely future WTO disciplines, has already signalled approval of this form of support and laid down criteria which must be observed if this immunity is to continue to be enjoyed. By embarking on the 'long game' of agri-environmental reform, lobbyists will find they have to situate themselves mentally in the rules, procedures and jurisprudence of the GATT if they are to ensure compliance with this new source of policymaking authority. Less clear, by comparison, is how environmentally beneficial the longer term WTO project of liberalizing agricultural policy will itself prove to be. A common assumption is that, by requiring the withdrawal of environmentally destructive price support, and releasing resources for an even greater expansion in agri-environmental spending than would otherwise take place, liberalization will yield a 'double dividend' for the rural environment of countries like the US and member states of the EU. Although an attractive idea that is well supported by theory, the double dividend thesis needs to be handled with care. Withdrawing price support, or even achieving substantial cuts in supported prices, is likely to have radically different agri-environmental implications for the managed countryside of the EU than in the farmed prairies of the US, for instance, while the 'green recoupling' of support may prove to be a politically more complicated process in both cases than is often assumed. Having spent most of the past decade lobbying for the greening of a basically protectionist set of agricultural policies and institutions, environmentalists are discovering that they need to engage in an entirely new debate, in which the abolition of the CAP and the US farm programmes is firmly on the agenda.

Agriculture in the GATT

The desire to liberalize agricultural trade has, of course, long been a feature of the US farm policy debate. As one of the world's great breadbaskets, the US retains a huge export potential which, unlike the EU, is based on genuine comparative advantage in the production of oilseed and grains (McMichael, 1993). Between 1962 and 1971, for instance, an average 46% of US wheat production, 13% of corn production and 31% of soybean production was sold on to world markets. These proportions had increased during the following decade to 58%, 27% and 39%, respectively (Runge, 1988). As was illustrated in Chapter 1, the history of US farm policy is punctuated with set piece battles between, on the one hand, free traders, keen to give American farmers the freedom to farm in order to capitalize on this competitive advantage, and New Deal interventionists, committed to the continuation of government programmes ostensibly designed to preserve the family farm (Benedict, 1953). After World War II, advocates of a more market-orientated farm policy appeared temporarily to gain the upper hand when academic economists, Republican party leaders and agri-businessmen reached a consensus about the need to 'take government out of agriculture'. In the event, the 1948 Agriculture Act went only part of the way towards this objective, carrying over wartime price supports but also requiring the Agriculture Secretary to adjust this downwards whenever supply exceeded what was deemed to be a 'normal' level (Cochrane and Runge, 1992). Ironically, it was very much at the behest of the US Congress that various escape clauses pertaining to agriculture were written into the original GATT Charter of 1947. These allowed signatories to continue supporting agriculture through the use of production subsidies and quantitative restrictions (Article XI) and export subsidies (Article XVI), throwing a *cordon sanitaire* around farm policy that would hold for the next 45 years. When the newly formed European Community came to set up its own agricultural policy in 1961, it exploited this loophole to establish one of the most protectionist support regimes in the world based on internal price support, variable import levies and, eventually, wider use of export subsidies. The US, meanwhile, had begun its drive for an export-led expansion of domestic agricultural production and was now persuaded of the need to bring agriculture within the GATT. It campaigned unsuccessfully during the 1963–1967 Kennedy Round to achieve full tarification of border protection measures and again under the Tokyo Round of 1973–1979. The Kennedy Round ended with a stand-off due to the EC's refusal to 'bind' (i.e. agree to legal limits on) variable import levies. Sheltered from world markets by the apparatus of the CAP, and encouraged to increase output by generous price guarantees, Europe's farmers now proceeded to turn a net EC import demand in 1963 into a substantial export balance. By 1992 the EC was exporting 27 million tonnes of subsidized grain on to world markets and had become the world's largest exporter of dairy products, meat and sugar (McCalla, 1993; Ufkes, 1993).

Witnessing this massive turnaround in market potential and the impact on world market prices of mounting surplus dumping, the US again campaigned unsuccessfully for farm support policies to be brought within the GATT under the Tokyo Round of 1973–1979. Fortunately for American farmers, the buoyancy of world markets throughout the 1970s meant that home producers, also underwritten by generous price guarantees, continued to enjoy an export-led boom, despite this increased competition from subsidized European exports. A sharp fall in commodity prices during the early 1980s, however, coupled with a loss in its own world market share, stiffened US resolve to put the liberalization of agricultural trade at the top of the agenda when the Uruguay Round was convened in 1986. According to Ronningen and Dixit (1991), the world teetered on the brink of trade war during these years, agricultural support being deployed to protect the incomes of farmers on an increasingly grand scale. Global expenditure on domestic farm programmes nearly doubled during the first half of the 1980s, and by 1986 the US and EU were each spending nearly $25 billion on agricultural support. Critics like Paarlberg (1989) were pointing to the futility of unilateral attempts to maintain US farm prices by paying farmers to take land out of production, observing that the 31 Mha taken out of production under the 1983 Payment in Kind programme had been met with a 25.5 Mha expansion in the production of foreign competitors. Meantime, the economic critique of agricultural protectionism was well advanced, with studies such as those by Anderson and Hyami (1986) quantifying the dead-weight loss imposed on taxpayers and consumers in industrial countries of 'beggar my neighbour' subsidization. These authors estimated that in 1986/7, 40% of the value of domestic support to US farmers merely offset the loss in profits caused by depressed world prices due to surplus dumping. Modelling studies designed to calculate the welfare gains of liberalization were called in aid of the argument for restoring market forces in agricultural markets. Indeed, quantification of the value of the transfers made to farmers through agricultural policy played an important role in bringing home to policymakers and opinion formers the extent of the burden imposed on consumers and taxpayers by agricultural, and particularly price, support, the OECD coming up with the concept of the 'Producer Subsidy Equivalent' (PSE), a percentage figure showing the value farmers receive in that country that is made up of government subsidies, to illustrate this (see Table 6.1). Wide variations in country PSEs, with the highly protectionist EU and Japan heading the list, supported the case of those pushing for a more level playing field in world agricultural trade. Moreover, the PSE analysis showed that subsidization was on an upward trend. The solution, it was widely agreed, was to end the use of coupled price support and export subsidies in order

Table 6.1. Net percentage PSEs on agricultural products, 1979–1986.
Source: Ingersent *et al.* (1994).

Country	1979	1980	1981	1982	1983	1984	1985	1986
Australia	7	9	10	15	11	10	14	16
Canada	26	25	26	28	28	33	39	49
EC (10)	40	35	31	34	33	33	43	52
Japan	68	61	58	62	66	67	69	76
New Zealand	15	16	23	27	36	18	23	33
United States	21	20	23	23	33	28	32	43

to close the gap between domestic and international prices, reduce the incentives for overproduction, and improve market access. Price support would be replaced by production-neutral, 'decoupled' income aids where continued government support could be justified on social grounds (Koester and Tangermann, 1977).

Government representatives appeared to concede the need for all this when they agreed in the opening Declaration to the Uruguay Round, 'to bring all measures affecting agricultural import access and export competition under strict and more effective GATT rules and disciplines' (GATT, 1986, quoted in Rayner *et al.*, 1993, p. 1514). According to Winters (1990) this was a revolutionary development, turning the focus on domestic agricultural support for the first time in the history of the GATT. Grant (1995, p. 8) predicts that the eventual result will be to open up the agricultural policy community by bringing in a wider range of actors with different priorities 'because not only trade ministers, but even heads of government may be drawn into the process if agricultural issues seem to be obstructing progress on broader questions'. In tabling its proposal for a 'zero option', which would involve eliminating all but fully decoupled subsidies to agriculture over a 10-year period, the US moved quickly to stake out the terms of a new debate about the international reform of agricultural policy. Its aim, and that of the Cairns Group of exporting nations which supported the zero option, was to reverse a decline in agricultural export earnings by curbing subsidized exports from the EU and elsewhere. The EU's stance was predictably more ambiguous, if not outrightly defensive. Its representatives resolved early on to continue the use of variable export levies and import subsidies in order to maintain a high domestic price regime and to uphold the principle of Community Preference. However, it was willing to concede the need to reduce the Aggregate Measure of Support (AMS) overall, provided that credit would be given for supply control. The tortured course of negotiations over the next 6 years is well documented (see Ingersent *et al.*, 1994; Swinbank and Tanner, 1996). The final URAA, reached in late 1993, was a considerable dilution, not just of this opening proposal but of the more moderate 'Dunkel Draft' tabled by the GATT Secretary General after talks had broken down in December 1991. Its significance though is that it commits GATT members to the progressive liberalization of agricultural policy through tarification, limits on the use of export subsidies and the decoupling of domestic support, and sets 1999 as the date for the commencement of a new round of negotiations, when these matters are expected to be further addressed.

Under the Agreement, there is to be a 20% reduction in the 'Aggregate Measure of Support' from a 1986–1988 base, an average 36% cut in border protection measures from a 1986–1988 base, a 36% cut in export subsidies from a 1986–1990 base and a 21% cut in subsidized export volumes, also from a 1986–1990 base. Commentators disagree about the immediate impact of these provisions on domestic agricultural policy reform. As Tangermann (1996) points out, the flexibility written into individual country implementation schedules, together with the choice of generous base periods for calculating the required amount of tarification, means that the new commitments do not bind signatories all that much, at least in the medium term. In addition, the accompanying Blair House accord between the US and the EU agreed certain compromises which further limit the immediate impact of these disciplines so far as domestic support is concerned. For the lifetime of this Peace Clause, payments offered to farmers for the purposes of production control or as compensation for price cuts (including the US deficiency payments) are to be classified as 'blue box' measures and thus exempt from the 20% AMS reduction requirement provided 'they do not grant support to a specific commodity in excess of that decided during the 1992 marketing year' (quoted in Swinbank, 1996, p. 397). More immediately constraining are likely to be the limits on the quantities of subsidized exports written into the Agreement. Increases in subsidized export volumes between 1986 and the beginning of the implementation period imply that significant cuts in these volumes will have to be made by some countries by 2000. Moreover, the 'end-loading' of many schedules means that these constraints will bite more deeply as the target date comes closer. At the same time, once the Peace Clause expires in 1999, the immunity extended to domestic support will be lost (or at least come up for renegotiation) and there is a general expectation (see Ingersent *et al.*, 1994) that pressure will be renewed to eliminate, or at least substantially reduce, all such partially decoupled production payments, as well as conventional price support, from national farm programmes. In Tangermann's (1996, p. 335) words, 'the rules of the game have been changed. The outcome of the game may not immediately be much influenced by the new rules, but the new rules will last, and in future they are likely to affect the outcome decisively'.

Decoupling by Degrees

In the US, this influence can already be discerned as policymakers seek to gain a negotiating advantage ahead of the next WTO Round. Market liberalizers like

Paarlberg (1989) seem to have won the day, at least on an intellectual level, with wide agreement that the US must put its own house in order if it is to press the case for further liberalization in 1999. As an influential OTA report put it 'not only are [existing domestic commodity programmes] detrimental in terms of trade opportunities and revenues lost; they also conflict with the spirit of international trade agreements, which the United States has, through the years, strongly supported' (OTA, 1995a, p. 7). Indeed, through a series of incremental policy reforms, policymakers have been quietly decoupling farm support from production for some time now. Previously, the 1985 Food Security Act had frozen the base yields used to calculate deficiency payment rates, effectively reducing the policy incentive for farmers to intensify production in order to increase the amount of subsidy received. Although not as radical as a freeze in base hectarage (which would remove the incentive for farmers to increase their hectarage of planted crops in order to increase future deficiency payments), the adjustment represented a significant first step towards decoupling support. More recently, the 1990 Food, Agriculture and Conservation Act gave farmers the freedom to plant non-subsidized crops on their base area without sacrificing their eligibility for deficiency payments at some future date. Known as 'zero certification', the idea was to increase farmers' ability to respond to world market conditions without suffering a policy penalty. Effectively, eligibility for deficiency payments was reduced to 85% of base area, farmers being free to plant any crop on the remaining 'flex acres' except fruit, vegetables or crops specially designated by the Secretary of Agriculture. Crops grown on this 15% of base area were still to be subsidized through price support and/or marketing loan prices, but these too were reduced – to 30% below the target prices which determine deficiency payments. As Zulauf et al. (1996) report, the effect has been to favour the replanting of cotton, corn and wheat and to discourage planting of sorghum, rice, barley and oats, giving an early indication of how selectively farmers would respond to the decoupling of price support.

   The 1996 Federal Agricultural Improvement and Reform (FAIR) Act takes this idea much further. This landmark legislation originated in a proposal by Senator Lugar of the Senate Agriculture Committee to eliminate nearly all agricultural support. By incorporating the proposal within the Congress Balanced Budget Act, Republicans hoped to move the new bill into law without the usual open-floor debate that would have been required with a farm bill (Orden et al., 1996). This tactic failed but the campaign for a 'Freedom to Farm' Act received fresh impetus in late 1995 when rising commodity prices began to soften the resistance of agriculturalists to the concept behind the bill. Pat Roberts, Chair of the House Agriculture Committee, saw a chance to progress his version of the original proposal by exploiting the projected decline in deficiency payments under current farm programme rules to obtain farm lobby support. Under his Freedom to Farm bill, a fixed hectarage subsidy would be paid to farmers each year from 1996 until 2002. These 'market transition' payments would be worth the same to a farmer regardless of the amount of land planted to crops or the yields obtained. To this extent, it was claimed, they were decoupled from production.

The effect has been to give American farmers more freedom to farm than for over half a century and to establish a benchmark against which agricultural policy reform in other industrial countries will be compared. Even so, FAIR owes its existence as much to political opportunism as to ideological commitment. One of the main reasons US policymakers apparently found it so easy to decouple their commodity programmes was the buoyancy of domestic market prices for many agricultural commodities throughout 1995 and 1996. The farm lobby saw its chance to wring a short-term windfall gain out of the policy process, exchanging deficiency payments that had by then lost much of their financial appeal for fixed land-based subsidies – in fact, Ayer (1996) demonstrates that high market prices had already neutralized the value of deficiency payments by the time the new arrangement was proposed, had the 1990 legislation remained in force. To quote Orden *et al.* (1996, p. 16), 'market prices in 1996 created an opportunity to end past farm programmes costlessly, ... the workings of the Congressional budget process yielding an otherwise unattainable change in farm policy'. Nevertheless, having made the change, policymakers are likely to find it hard to reverse (President Clinton's election campaign commitment to recouple some farm payments to production in order to assist some of the more vulnerable family farms notwithstanding). While it is true, as Harvey (1996) indicates, that FAIR does not rule out a return to traditional forms of support – the base legislation having been suspended, not annulled – it is hard to conceive the Administration forgoing the WTO negotiating advantage of having an already decoupled farm policy by reinstating deficiency payments. Margaret Tutwiler, chief US negotiator at the WTO talks, has already claimed that FAIR puts most of the support for US farmers comfortably within the green box, and has called on the EU to follow suit (Tutwiler, 1996). To the extent that it is the culmination of a decoupling process that has been in progress since 1985, FAIR also reflects a shift in attitudes among US policymakers and the farm lobby in favour of a more market and export orientated agricultural industry (Offutt, 1996). A survey of farming opinion conducted by Martin *et al.* (quoted in Orden *et al.*, 1996), shows, for instance, that fully 41% of a sample of 10,000 farmers in 15 key agricultural states now favour a phasing out of agricultural support compared with just 23% in 1980 and 35% in 1990. In the wake of FAIR this now begins to look like a realistic long-term possibility.

For European policymakers, decoupling and liberalizing the CAP was always going to be a politically more difficult project. The status of the CAP as the EU's only truly common policy means that attacks on it have traditionally been construed as *non communautaire*, and thus tantamount to an attack on the European integration project as a whole. Among other things, liberalization implicitly challenges the principle of Community Preference which is at the heart of the CAP and makes it more difficult to preserve a Green Europe of small farms through partially decoupled payments. It was European reluctance to offer more than a 30% cut in internal support and a refusal to table concrete proposals for reducing export subsidies, which initially stalled the GATT talks in 1990 (Swinbank and Tanner, 1996). Commentators disagree about the extent to which this breakdown put pressure on farm ministers to agree the 1992 MacSharry reforms (see Josling, 1994). According to Tangermann (1996), the Agriculture Commissioner used the GATT talks to lever ministers into accepting changes to the CAP that were needed anyway on domestic budgetary grounds. For political reasons, the sequence had to be CAP reform first and a GATT agreement second – for otherwise ministers would have found themselves forced into overhauling the CAP solely to fulfil international obligations. Central to the proposals tabled by Agriculture Commissioner MacSharry in 1991 was a cut in the level of price support, to be offset by a system of hectarage (and in the case of livestock producers, headage) compensation payments (CEC, 1991a). While the MacSharry Plan, as it became known, went only part of the way towards meeting the demands of the EU's trading partners (tarification was not mentioned and the proposed compensation payments were only partially decoupled from production), it created a basis for negotiation.

In its final form, the CAP reform package agreed in May 1992 brought about a 36% cut in cereal and beef support prices. In the cereal sector the Arable Area Payments Scheme offers farmers compensation payments for price cuts calculated on the basis of historical cropping patterns and average yields. Larger producers must set aside part of their cropped land in order to qualify for this compensation. Beef producers also qualify for compensation payments calculated on a headage basis, while the headage payment system for sheep and goats was revised in order that these too would qualify as decoupled payments under the Agreement. (All livestock subsidies were made subject to quotas, ceilings and stocking-rate limits, mechanisms designed to limit the policy incentive for livestock farmers to maximize livestock numbers in order to receive more payments.) Apart from a modest cut in milk quota, the dairy regime was left more or less unchanged, as was the support policy for sugar. These significant lacunae notwithstanding, the effect was to improve the EU's negotiating position because it could now claim the CAP had been put on a liberalizing path. Reductions in the volume of subsidized exports were now in prospect because of partial decoupling and set aside, while reduced threshold prices had improved market access – all measures calculated to win the approval of the US and the Cairns Group. As a result of the MacSharry reforms, in fact, the AMS for cereals, exclusive of the compensation payments, was reduced significantly and under the Agreement that was later signed could be offset against more modest reductions in other sectors. (There is an argument that without this reduction the EU would have later found itself making far deeper cuts in price support in order to meet its URAA commitments, assuming an Agreement would have been concluded at all.) The 1992 reforms also allowed the EU to more easily meet its eventual URAA commitments on export subsidization. Cuts in price support combined with set aside meant a significant decline in subsidized export volumes compared with EU projections (Harvey, 1995). Further barriers to agreement were cleared when the Blair House Accord recognized the MacSharry compensation payments as notionally decoupled, and thus exempt from the AMS reduction requirement. In Swinbank's (1996) judgement, the URAA which followed was the maximum EU farm ministers were willing to concede, and the minimum the US and other trading partners were prepared to accept. Significantly, the MacSharry reforms had a less immediate impact on the EU's farm budget. Far from cutting FEOGA expenditure, in fact, the reforms added to it because of the cost of the new compensation and other direct payments. Between 1993 and 1995 FEOGA expenditure on cereals and set aside increased by 5.7 billion ECU. Tangermann (1996) estimates that the total net cost of the reforms to the Commission in the cereals sector could be over 8 billion ECU, to which must be added a further 2 billion ECU expended on the Accompanying Measures. For this author, the fact that the 1992 reforms increased rather than reduced expenditure adds weight to the argument that they were primarily motivated by the need to make the CAP consistent with the Agreement likely to emerge out of the WTO.

Even so, few commentators believe that this is the end of the affair and there has been much speculation about the reverse impact of the URAA on the CAP. Most commentators assume that the compensation payments will lose their immunity after 1999, when the URAA is up for renegotiation. At this point the EU will be required to honour its commitment to reduce the AMS through real cuts in the level of domestic support unless it can come up with a system of direct payments that is more acceptable to the WTO and its trading partners (see below). In Harvey's (1995, p. 210) opinion 'the passing of the present compensation arrangements as within the green box (*sic*), and thus non-trade distorting, is widely understood to be a convenient fiction for the purposes of the current Agreement only'. US officials, confident that their own FAIR subsidies will be granted a permanent green box classification, are already pointing to the anomalous status of the EU's compensatory payments (which in 1995/6 cost the EC 6.6 billion ECU) in WTO terms and calling for their abolition (Tutwiler, 1996). Meanwhile, the export subsidy constraints on the EU built into the URAA are likely to have begun to bite. At present, a rise in world market prices combined with the domestic use of set aside mean that the EU remains comfortably within the limits on the volumes of subsidized exports allowed by the Agreement (Swinbank and Tanner, 1996). A continuing improvement in agricultural productivity, together with the fact that the URAA export constraints are 'end loaded', becoming more restrictive at the end of the implementation period than at the beginning, suggests that these will become more binding as time goes on. While the EU may allow intervention stocks to build up, this would increase the budgetary costs of the CAP and is unlikely to be politically acceptable for long. Moreover, the likely accession of the Central and Eastern European (CEE) countries at the turn of the century will bring with it a large addition to the EU's productive potential but only modest additional allowances for subsidized export volumes under the URAA (Tangermann and Josling, 1995; Thomson, 1996).

In any event, with their huge potential comparative advantage in the production of many agricultural products, it is thought that the CEEs would respond to price guarantees presently available under the CAP by expanding output and quickly putting strain on the CAP's system of market management. Few believe that the CAP could sustain the resulting increase in expenditure on export refunds and market intervention, even if they were allowable under WTO rules. As a UK Foreign Office minister put it: 'I think it would be impossible on budgetary grounds to bring the CEEs within the CAP in its present form. Put crudely, I think it would bankrupt the European budget' (quoted in House of Lords, 1994, p. 122). The verdict of most policy experts is that:

> the problems become more acute the further forward one looks, as consumption levels are unlikely to show significant increase and output, fuelled by productivity gains and sustained by CAP price support, is likely to continue its increase. The situation threatens to become unsustainable if, by the early 2000s, the EU has embarked upon another enlargement to embrace states of Central and Eastern Europe.
>
> (Swinbank, 1996, p. 403)

It is at this point that many commentators expect the further scaling back of price support to occur if the EU is to fulfil its international obligations and remain within budgetary limits. To quote Swinbank (1996, p. 407) again: 'The Agreement arranged the straitjacket around the CAP. The next Round will tighten the cords and begin the painful process of dismantling CAP support'.

The Double Dividend

If this occurs, European farmers will join their American counterparts in experiencing one of the most decisive reductions in agricultural support for 40 years. The environmental consequences could be profound. For some commentators (Jenkins, 1990; Anderson, 1992), the withdrawal of conventional price support, already under way in the US and now a long-term prospect for the EU, promises to be doubly beneficial for rural environments. To begin with, a reduction in production subsidies should bring about a relocation and extensification of agricultural production and reduced pressure on the conservation resource. Additionally, there should be scope for reallocating the resources previously tied up in price support in favour of decoupled environmental programmes. This thesis has strong theoretical support. Globally, it is expected that there would be a relocation of production away from the industrial countries in favour of developing countries, with the biggest declines in grain and meat production taking place in Japan and Western Europe, offset to some extent by an expansion in Australia and North America (Anderson and Strutt, 1996). This is predicted to be environmentally beneficial because, while chemical and fertilizer use would undoubtedly expand in developing countries, this would be from a low base, and should be more than neutralized by absolute declines in chemical use from a very much higher base elsewhere (Anderson and Strutt, for instance, point out that, on average, industrial countries use ten times more fertilizer and pesticides per hectare than Australia and developing countries). Regarding the extensification effect itself, the argument from theory is that the withdrawal, or substantial scaling down, of commodity price support in industrial countries will reduce the incentive for farmers to use 'land saving' technologies and farming inputs. This is because the benefits and expectations surrounding government price support have in the past been capitalized in land values. Cutting price support will depress land values and rents, making it less necessary for farmers to economize on their use of the land input by applying more fertilizers and pesticides to each hectare of crops or by putting more stock on every hectare of grass. It is widely assumed (see Harvey and Whitby, 1988; Jenkins, 1990; Abler and Shortle, 1992) that the resulting extensification of production will be good for the environment, reducing surface water and groundwater pollution, eliminating over-grazing and easing the pressure to reclaim land from the wild.

As for the 'green recoupling' of support which is supposed to accompany this, the economic theory of public goods identifies a case for continued government intervention where there are environmental services that would be underprovided if left to the market. In the agricultural case, there is a further, more pragmatic argument for redirecting support into environmental schemes. According to an evolutionary model of farm policy change, the decoupling process, once begun, puts mounting pressure on policymakers to find new ways of justifying agricultural support. This is because substituting income compensation payments for price support increases the visibility and transparency of government subsidies, shifting the burden from consumers to taxpayers and engendering a debate about why farmers should continue receiving public support. As Rausser and Irwin (1989, p. 363) observe, 'consumers are generally less vocal about policies that affect them only slightly, while the taxpayers' burden is more obvious not only to the taxpayer but also to his elected representatives who impose taxes and authorise expenditures'. Given that paying farmers to produce environmental goods is likely to be a more publicly defensible position than supporting them simply because they are farmers, it might be expected that over time more and more expenditure would be switched into green subsidies. In support of this, advocates of the double dividend can point to immunities already enjoyed by green payments under the URAA. By placing environmental subsidies within the 'green box' (the colour is coincidental), negotiators signalled recognition of their likely future importance as a non-trade distorting source of support. They also of

course encourage governments to begin to substitute them for the more con-
troversial blue box compensation payments which currently absorb a much
greater share of agricultural spending.

The idea of a double dividend is thus extremely compelling, not least
because it suggests that improving agricultural competitiveness and protecting the environment can be complementary policy goals (Potter, 1996). In
the US especially, this has considerable appeal to policymakers attempting to
square demands from the farm lobby to be given 'the freedom to farm' with
public concern about the environmental consequences of export-led agricultural expansion (see, for instance, Creason and Runge, 1990; OTA, 1995a;
1995b). In practice, though, considerable uncertainty surrounds both sides
of the double dividend equation, particularly in a European context, where the
environmental benefit of withdrawing support could prove less decisive, and
the green recoupling of support harder to achieve politically, than is often
assumed. Each is worth further analysis, if only to identify some of the choices
policymakers may have to face in dealing with the environmental consequences of the liberalization of agricultural trade.

A great deal rides on the extensification of production which cuts in
commodity price support are supposed to bring about. The assumption is that
farmers will respond to the induced change in relative factor prices by making
inroads into fertilizer and pesticide use that are sufficient to have an environmental effect. According to estimates presented by Abler and Shortle (1992),
for instance, a reduction in variable input use will be by far the strongest short-
run reaction to a substantial cut in commodity support. Their estimates for the
EU suggest a 60–86% medium- to long-run cut in fertilizer and chemical use
on cereal farms following the decoupling of price support under the CAP. As
Harold and Runge (1993) point out, however, the elasticities on which these
projections rest have weak empirical support and are in need of further research before really safe predictions can be made. Crabtree (1992) observes
that current estimates of the elasticity of demand for inputs with respect
to product prices vary widely, with programming models usually assuming
fertilizer use to be less price sensitive than econometric ones. This does not
instil much confidence in their accuracy. Tobey and Reinert (1991) discovered
that reducing the estimated elasticity of substitution between fertilizer use and
land significantly reduced the projected environmental benefits in the US of a
reduction in commodity support. For the EU, Leuck and Haley (1996) estimate
that large reductions in livestock densities and fertilizer use would have to
occur if CAP reform is to bring about an extensification of production large
enough to meet the targets laid down in the EU's own Nitrates Directive. Even
assuming input use falls significantly, the precise nature of the environmental
benefit being predicted is rather unclear. Much depends on the starting point
from which a reduction in, say, fertilizer use is being made and on the environmental vulnerability of the locality concerned. The Dutch farmer who reduces
his fertilizer application rate from 700 kg ha^{-1} (the country average for The

Netherlands in 1992) to 500 kg ha^{-1} will have achieved a substantial cut in fertilizer use but is unlikely to have done enough to eliminate environmental damage due to runoff or leaching if the rate of uptake by the crop is only 150 kg ha^{-1} (Harold, 1992). Burt *et al.* (1993) point out that a reduction in nitrogen use (for example) does not automatically mean an improvement in environmental quality as it is losses to the environment through leaching and runoff that determine whether any pollution damage actually occurs. In any case, the real environmental impact of an input reduction critically depends on where it takes place, 'the way in which pollutants move through the soil and hydrological system [determining] the damage caused' (Hodge, 1992, p. 68). Baldock and Beaufoy (1992a) argue that if the aim is to improve groundwater quality by reducing nitrate pollution, for example, then targeting selected farms in environmentally vulnerable locations for substantial reductions is likely to prove more effective than the wide but shallow extensification of production that will be triggered by price cuts.

Further complicating the picture are farmers' decisions about how much land to keep in production and what to grow on it. In the US, the FAIR reforms have already brought more cropland into production as land previously set aside under the ARPs has been returned to cultivation by producers no longer required to 'cross-comply'. It is estimated that cropland will expand by between 1 and 3% after full implementation of the North American Free Trade Agreement (NAFTA) and the URAA (Ervin and Keller, 1996). It is further predicted that the return of all 6 Mha previously idled under these programmes will be needed to meet the additional US export demand by 2000 (OTA, 1995a). There is little evidence to suggest that conservation practices will be maintained on this land without the incentives government programmes provide (Lovejoy and Napier, 1986). If liberalization is still to produce an environmentally beneficial outcome, the extensification effect will consequently have to be great enough to offset the damage caused by this and any further conversion of cropland that takes place. Most analysts conclude that the outcome is likely to be finely balanced. According to Miranowski *et al.* (1991) total soil erosion can be expected to increase because of the larger area of land under crops, though chemical use should decline due to the effects of lower prices and changes in the mix of crops grown on farms. If CRP lands were also returned to production, both erosion and chemical use would increase. This study suggests that a unilateral liberalization of US farm policy would, over the medium term, result in some 33 million tonnes of additional soil loss due to erosion but a cut in fertilizer use on corn of 420,000 tonnes. In Tobey and Reinert's (1991) study, the substitutability of ARP land for chemical use emerged as the key determinant of the environmental impact of policy reform. Again, the benefits of lower farm chemical use were thought likely to just outweigh the costs of increased erosion due to more land being brought into production. Overall, though, the magnitude of the effects was small. A simultaneous 40% conversion of ARP land and a 40% cut in producer support was predicted to result in a modest 8.5% reduction in the environmental damage costs of US agriculture.

The interest of Europeans, by contrast, is much more likely to focus on the effects of land abandonment and the restructuring of agricultural production on the extensive margin. In this more managed countryside, the environmental consequences of dismantling support are likely to be far less neutral than they would be in the US, leading commentators some time ago to conclude that 'while high prices have contributed to environmental damage in the past, merely reducing them could yield a variety of effects – not all of them beneficial for the environment' (Haigh and Grove-White, 1985, p. 7). Two observations can be made. First, there is likely to be a relationship between the pace of reform and the nature of the associated environmental impact (this relationship obviously applies equally to the US, though with arguably less critical implications for landscape quality or farmland habitats). Rapid and abrupt price changes will give rise to the most dramatic changes in the ecology of the European countryside and its character as falling land and asset values drive some farmers out of business and bring about a redistribution of land holdings and a change in the pattern of land management. A phased reduction in support would give farmers more time to adjust and may be associated with rather more lags in the adjustment process. Land use and environmental change is now more likely to be the cumulative result of decisions made by farmers who can no longer earn satisfactory returns from agriculture. A range of land use changes will occur because of enterprise substitutions and business restructuring. In both cases, however, the logical outcome is a smaller industry of fewer, larger businesses because the CAP 'brake' will be off and farmers will be under tremendous pressure to spread fixed costs by farming more land. Early victims of a rapid draw-down of support would be highly geared farmers, including those who have borrowed money to finance expansion. Colman (1983) argues that, by encouraging heavy borrowing against the collateral of inflated land values, high price support has created a policy trap which has ensnared precisely those most dynamic farmers who have undertaken capital investment to expand production, particularly of cereals and milk. Falling prices and asset values will also eventually remove many other categories of farmer who are simply marginal in economic terms. Some of these will own or manage land and embody skills essential for the conservation of the countryside and, to the extent that many remaining habitat fragments in the lowland countryside are the result of accidents of land occupancy or family history (Potter and Lobley, 1996), this marginalization process will further deplete the conservation resource to be found there.

Indeed, a very European concern will be whether there will be sufficient resources, skills or farmers in a post-GATT world to ensure that wildlife on farms, larger tracts of semi-natural vegetation and the cultural fabric of the countryside are adequately maintained. Thus, while a decline in farm profits and land values may well ease the pressure to reclaim habitat, drain wetland or remove hedgerows (Harvey and Whitby, 1988; Harvey, 1990), it is less clear what the implications will be for the more active management of wildlife habitat on farms or for the protection of countryside character. Commentators such as Bowers (1995) argue that the visual quality of many of Europe's agricultural landscapes depends on the continuation of farming and the management and upkeep of components like hedges and drystone walls, and vegetation such as moor, heath and woodland. For Webster and Felton (1993) undermanagement is arguably already as great a threat to the nature conservation interest of large parts of the farmed countryside, and work conducted by McCracken and Bignal (1995) has drawn attention to the likely disappearance of high natural value farming systems of various sorts should agricultural support be withdrawn. By definition, such systems are already economically marginal, employing 'practices which have been out of fashion for many years and techniques which are not generally part of modern agriculture' (McCracken and Bignal, 1995, p. 31). Other things being equal, the withdrawal of support can be expected to put their survival at risk, for, as Bowers (1995, p. 1238) comments, 'there is no reason to suppose that these less intensive techniques would be profitable at market clearing prices' given their marginal profitability at current supported prices. In the Causse Mejan in France, for example, traditional, low intensity producers are already responding to the partial decoupling of support by changing their mix of enterprises and increasing output of sheep's milk in order to produce Roquefort cheese. Beaufoy *et al.* (1994) report that farm amalgamations are increasing, with a greater likelihood that the most marginal land is abandoned or farmed more extensively. With this comes a series of ecological changes as species-rich grassland is replaced by coarse herbaceous vegetation, scrub and, eventually, high forest. Similarly in the Black Forest region of Baden-Württemberg, the continued marginalization of extensive cattle production, a process expected to accelerate with uncompensated liberalization, is having a serious impact on the landscape and its conservation value, colonization by scrub and the opportunistic planting of pine by absentee farmers, creating habitats of lower value than the grassland vegetation they replace (Luick, 1996). In Sweden, Vail *et al.* (1994) report a significant reversion of farmland to forest due to a fall in agricultural returns following that country's partial liberalization of its agricultural policy in 1990. In southern member states, one of the predicted effects of liberalization is an increased risk of exposure to soil erosion and fire as terracing practices decline and understocking exacerbates the fire hazard by encouraging the invasion of scrub on grazing land (Beaufoy *et al.*, 1994).

A second prediction is that the environmental impact of the removal of support will be irregular and spatially uneven given the diversity of farming situations with which policy change has to interact. The geography of the farmer response is consequently very important in the overall picture. If the general forecasts made by researchers such as those based in The Netherlands Scientific Council for Government Policy (NSCGP) are right, liberalization will bring about a general contraction in Europe's farm output as world production gravitates to countries and regions with greater comparative advantage, particularly in dairy and cereals (NSCGP, 1992; Harrison *et al.*, 1995). Although some European cereal production would remain competitive at world market prices, this is likely to be restricted to the arable heartlands, leading to a significant relocation of production at a European scale. Under a free-trade scenario, in fact, agricultural activity is expected to concentrate in the northwest of the EU, arable farming becoming confined to southern Germany, eastern England and eastern and western France (NSCGP, 1992). Some grass-based livestock production in northern member states will also remain competitive, as will intensive, grain-based systems of pig and poultry production. The main casualties are expected to be dairying, olive oil and wine production, all of which would contract sharply following liberalization. For a member state like the UK, the drawing back of arable production to the best land in East Anglia, the East Midlands and parts of the South would have a variety of environmental effects, the most obvious of which would be a shift out of tillage and into grass. Existing grassland farmers on poorer land could find themselves squeezed by an expansion in livestock production on dairy farms and on traditional mixed farms. Gould Consultants (1986) predict the emergence of new marginal lowland countryside where the effects of this shunting of production would be most acutely felt. In such 'middle countryside' land will revert to grass but at varying intensities of use and where physical conditions allow there will be a tendency for land to move out of agriculture altogether and into commercial forestry. Lowe *et al.* (1995) expect that it will be movements of land out of farming and into forestry and industrial crops, rather than the smooth changes in farming practices that have been seen in the past, which will increasingly become the main focus of environmental concern. According to the Gould Consultants (1986) study, these are likely to be most pronounced in the remoter LFAs, where the withdrawal of support, even if phased over several years, would have dire consequences for the occupancy and management of land. Farm businesses already 'on the edge' economically would likely disappear before enterprise adjustments could be made, and the land released may be bought up at bargain prices by neighbouring farmers, private forestry companies or, conceivably, conservation organizations. Amalgamation could bring with it extensification of a sort few conservationists desire, leading to the ranching of large tracts of upland vegetation and a decline in the management practices essential for sustaining biodiversity and ensuring landscape protection.

Green Recoupling

The case for green recoupling, it would seem, is very strong if farmers are to be provided with sufficient incentive and resources to continue conserving soil, preventing pollution and managing the countryside after conventional forms of support have been withdrawn. In a European context particularly, the environmental and nature conservation effects of significantly reducing price support do not run in the same direction, suggesting a strong need for counter-balancing measures which would be deployed to offset, and not merely complement or speed up, some of the first round repercussions of liberalization. The case for special assistance to keep high natural value farming systems in place would seem to be very well made, for it could (and will) be argued that without this, liberalization will wipe out much of the human capital necessary for the effective conservation of the European countryside. The UK's Country Landowners' Association (CLA, 1994, para. 165), for instance, recently opined that 'concerted efforts will need to be made as CAP reform proceeds to prevent the loss of sympathetic land management systems through intensification and abandonment ... These systems will need to be protected, sustained and encouraged'. In Spain, the Ministry of Agriculture has identified stable rural populations as a policy priority, while the European Network of Alliances for Sustainable Development argues that 'the best guarantee for environmental conservation is the maintenance of the population that [already] manages the land' (SAFE, 1996, p. 31). There will also continue to be calls for government action to tackle a countryside management and farm pollution problem likely to be only marginally affected by the shallow extensification of production induced by price cuts alone. In short, much more would seem to be required of policymakers after price support has been withdrawn than a straightforward reading of the extensification effect might suggest. Obtaining a favourable environmental outcome from the liberalization of agricultural policy will depend heavily on the ability and willingness of policymakers to divert sufficient resources into agri-environmental programmes to ensure the continued management of the farmed landscape and the environmental capital it contains. As Kuch and Reichelderfer (1992, p. 224) conclude from their review of the situation in the US, the liberalization of agricultural policy 'is an exceptionally inefficient approach to enhancing environmental protection. The superior approach [is to] implement environmental policies or programmes directed specifically at the problem'.

Advocates of the double dividend appear confident that policymakers will be willing to do this, though they concede that much depends on the ability of environmentalists to articulate the case for investing substantially greater sums of public money in schemes that are often still at an early stage of development (Jenkins, 1990). Supporters of this approach have in mind a large expansion and refinement of agri-environmental policy, in which targeted environmental payments to farmers are fully decoupled from both production and income support. According to Jenkins (1990, p. 7) 'expenditure on such supports can only be justified if they ensure the provision of specific environmental goods'. Buckwell (1989, p. 159) further argues that 'in each case, it is not only necessary to identify targets but also to define criteria for monitoring the success or otherwise of policy changes designed to achieve specific targets'. Theory would suggest that the assumed income boost from liberalization will increase demand for environmental quality improvements amongst US and EU citizens. This is a version of the argument that free trade is good for the environment because it generates the economic growth which increases the demand for environmental protection as well as providing the resources necessary to achieve it (see Ekin *et al.*, 1994). Recent reviews of the evidence on the relationship between income growth and environmental concern, however, suggest that the income elasticity of environmental improvement may be less than one (OTA, 1995a). So far as the demand for agri-environmental services specifically is concerned, Whitby and Saunders (1994) point out that there is presently no objective measure of taxpayers' willingness to pay for an expansion of agri-environmental programmes and no necessary reason why, in aggregate, this should equal existing levels of agricultural support. The main sources of uncertainty are political. To what extent will taxpayers be prepared to pay for such an expansion in agri-environmental support, particularly when the benefits may be difficult to quantify and slow to appear? How far will policymakers be able to defend these new environmental policies from rent seeking by agricultural interest groups intent on preserving their traditional policy entitlements in what will be a period of farm policy retrenchment?

These questions have been exercising American agri-environmentalists for some time. Experience with the CRP suggests that they are intimately connected because rent seeking erodes environmental performance and undermines public confidence in, and willingness to pay for, the programme itself. On the other hand, the state of knowledge about agri-environmental problems in the US is now sufficiently good to allow policymakers to target green payments so that the risk of rent seeking is much reduced (OTA, 1995b). As Babcock (1996) points out, more than 98% of the improvement in water quality being effected by the CRP in 1995 could be maintained by selectively enrolling just 27% of the land. With access to raw data from the NRI and high quality analyses of alternative policy options and their environmental effects produced by its own Economic Research Service (ERS), together with information contained in the next National Soil Conservation Plan due to be drawn up by SCS in 1997, USDA is now well placed to 'micromanage' its programmes in order to improve their environmental performance (Zinn, 1994). The consensus during the run up to the 1984 Farm Bill was that, if the green recoupling of US farm policy was to take place, it would have to be achieved through measures that were carefully designed to provide the broadest and most enduring environmental benefits for every dollar spent. In its briefing for the 1984 Farm Bill, the AFT (1996) acknowledged that the policy process was overwhelmingly driven by the need to reduce the size of the federal deficit. 'Federal rollback', as it is called, a deep-seated trend that has been under way since the early 1980s, makes it unlikely that the conservation dividend from liberalization will be all that large.

Pressure to cut federal expenditure makes it more important than ever that soil conservation programmes are reformed so that they are more efficient in meeting policy objectives. Green recoupling would also need to be compatible with URAA rules which require that green payments are, first, part of a clearly defined government programme, second, have no or minimal trade-distorting effects, and third, are limited to subsidizing the added cost or lost income from the practice adopted or technology shift accomplished. Although the idea of replacing the commodity programmes with a system of green payments was proposed and discussed during the 1996 Farm Bill process (see AFT, 1996), the geographical redistribution of support that would be necessary to ensure this was achieved efficiently proved too politically sensitive for it to enjoy wide support. Instead, Congress recommissioned the CRP and expanded the WRP, effectively reinvesting some of the dividend from the abolition of deficiency payments in existing policy programmes. Under FAIR, the federal government undertakes to fund a CRP up to a maximum of 14.2 Mha, though the Secretary of Agriculture is also empowered to terminate contracts that are at least 5 years old and to target for re-enrolment environmentally sensitive land. The government is also required to set up an Environmental Quality Incentives Program (EQIP) to provide technical, educational and cost-sharing assistance to farmers. Whether FAIR is the 'very pro-environmental bill' proclaimed by Zulauf *et al.* (1996) remains to be seen. These authors point to the popularity of the new market transition payments with farmers, who are discovering they no longer have to set land aside in order to qualify for a government cheque. To the extent that conservation compliance and sodbuster provisions still apply to the new decoupled payments, environmental leverage has thus been increased. It would however be too much to claim, as do Zulauf and colleagues, that this effectively makes the transition subsidies 'green payments', because the compliance conditions appear to have been relaxed even as they have been more widely applied (Ayer, 1996). The reforms do nevertheless give some encouragement to the view that green re-coupling can be the corollary of liberalization. In the meantime, the $35 million allocated to the EQIP looks to some suspiciously like a small crumb from the budget cutters' table.

The debate is moving on. Many conservationists are acknowledging that the return of large areas of US farmland to cropping is inevitable. A strategy of relying exclusively on subsidized land diversion to address soil erosion and water pollution is increasingly seen as no longer viable on either free trade or budgetary grounds. Other approaches, including the promotion of technical and agronomic solutions through cost sharing appear to be required, preserving US farmers' freedom to farm while minimizing environmental costs. This connects with a view of the causes of soil erosion and agricultural pollution which puts much more emphasis on the structure of property rights and technical change than on policy influences alone (Ervin and Graffy, 1996). What now seems to be emerging is a bi-modal approach to agri-environmental policy design, or at least its advocacy by many commentators and lobbyists. On the one hand, better targeted but spatially less ambitious land retirement programmes, micro-managed from the centre to maximize environmental effectiveness; on the other, a harnessing of the profit motive and an appeal to the stewardship of individual farmers to ease the environmental problems associated with the larger areas of cropland which look set to remain in production. According to an influential report from the OTA (1995b), for example, there is scope both for better targeting of federal programmes and for encouraging farmers to adopt 'complementary technologies' such as conservation tillage and integrated pest management which 'by design, enable farmers to enhance environmental quality while maintaining farm productivity and profitability' (p. 33). This analysis drew on a Delphian exercise to review the latest scientific evidence concerning the incidence and magnitude of agri-environmental programmes in the US and to map a new set of priorities for public policy and research. The study concluded that long-term land retirement was only cost effective where agricultural production was fundamentally incompatible with environmental protection and that farmers should instead be encouraged to adopt management strategies and technologies on a broader front. As the report put it:

> Bridging the gap between current programmes and new realities will involve more refined targeting plus, in response to calls for streamlining programmes to reduce burdens on farmers, a simplified set of approaches that take maximum advantage of private incentives (and rely less on subsidies).
>
> (OTA, 1995b, p. 32)

The implication was that at least part of the conservation dividend should be invested in the research and development of such technologies, and the rest on subsidizing their uptake on farms, it was implied.

If this happens, it will not take place in a legislative vacuum because, paradoxically, federal rollback and agricultural liberalization make it more likely that the states will move in to regulate farmers in order to meet stricter environmental quality standards for soil, water and wildlife. As Ringquist (1993), quoted in Ervin and Graffy (1996), discovered, the 1980s saw the enactment of state environmental regulations that were frequently much more rigorous than federal rules. So far as agriculture is concerned, minimum environmental standards are already being laid down which legally oblige farmers to maintain prescribed levels of water quality and soil conservation. In Vermont, for instance, a uniform code of good agricultural practice has been agreed which defines a safe minimum for all farms (OTA, 1995a). The public appetite for such standards appears to be growing, according to a survey conducted by Hoban and Clifford (1994). These researchers discovered that while the majority of respondents continued to blame industry for water pollution, 'farmers may face an increasingly restive public as the environmental consequences of many present-day farm practices are measured and recognised'

(Hoban and Clifford, 1994, p. 168). If acted on, this trend towards more regulation may redefine the situations in which green payments can be offered to farmers, limiting compensation to cases where improvements are being undertaken over and above the standard and forcing the further adoption of complementary technologies.

European policymakers face a rather different set of challenges in their attempts to recouple agricultural support to environmental policy goals. If recent history is any guide, there will be much greater pressure here to conflate environmental protection with the support of farmers' incomes in the way environmental programmes are conceived, designed and delivered. Critics argue that this may compromise their environmental performance and thus their future defensibility as public expenditure sinks. The complication, as was demonstrated in Chapter 5, is that the agri-environmental problem is often defined in ways which blur the line between farm income support and environmental protection and so make a fully decoupled policy harder to achieve. The case for special assistance to keep high natural value farming systems in place will be strongly made; indeed, the risk of desertification and its undesirable socio-environmental effects is already emerging as one of the main justifications for a defensive expansion in agri-environmental spending after the next WTO Round (Buckwell, 1996). Whether this is a defensible use of the conservation dividend depends on the view taken of desertification and agricultural decline as an environmental threat. The argument in its favour is that the public values cultural landscapes that are the result of widely flung patterns of agricultural occupancy and activity. It is thus legitimate to seek to keep certain types of farmers in place, probably through a system of permanent hectarage payments, to which environmental conditions might later be attached. Hodge (1992, p. 71) comments that, in these terms, the support of farmers' incomes and protecting the environment are inextricably linked 'such that proposals for reducing prices appear inescapably to imply a denial of environmental values'. There is little to suggest that this policy doctrine will change. Elements of it were on display at an important conference convened by the European Commission in 1996 to debate the future shape of European rural policy (CEC, 1996a). In a reassertion of Green Europe (some would say agricultural fundamentalist) thinking, the Irish Minister for rural development declared that 'agriculture continues to be the lifeblood of rural areas ... the pressures on agriculture should be used to encourage policymakers to formulate measures that ensure the maintenance of the maximum number of farm families' (Deenihan, 1996, p. 7). The Commissioner for Agriculture amplified this when he argued that 'rural society is a socioeconomic model in its own right which must be preserved in the interests of European society as a whole' (Fischler, 1996, p. 12). And in the wider sphere of debate, commentators like Berlan-Darque and Teherani-Kraner (1992) and Baudry and Laurent (1993) have detected continuing popular support for policies which are designed to maintain the perceived link between farmers, the environment and rural society.

According to the standard formulation, however, environmental payments must be fully decoupled from both production and the support of farmers' incomes if the capture of environmental programmes by rent seeking agricultural interest groups is not to occur (Rausser and Irwin, 1988). Jenkins (1990, p. 7) puts it like this: 'expenditure on such supports can only be justified if they ensure the provision of specific environmental goods'. The UK House of Lords Select Committee on the European Communities (House of Lords, 1991, p. 48) further specifies that 'payments, which should not be restricted to farmers alone, should be graduated according to the environmental interest of the land, the sophistication of the management demanded or the resultant changes in farming practice'. Under what this author has called a 'radically decoupled' policy (Potter, 1996), conventional agricultural support would be withdrawn, to be replaced by a lightly engineered system of environmental payments, targeted to maximize environmental value for money (see Table 6.2). Such an arrangement would seem at first sight to be the logical fulfilment of past agri-environmental reform in the EU because, unlike the existing 'weakly decoupled' policy, there would be no competition between production support and environmental incentives and no need to resort to environmental accommodations such as conservation compliance in order to promote conservation on farms. In reality, radical decoupling has a number of drawbacks, chief of which would be the inevitable pressure to sacrifice the breadth of coverage of schemes in order to concentrate resources in the hands of a selected number of farmers best qualified to produce the specified environmental

Table 6.2. Environmental aspects of decoupling. Source: Potter (1996).

Action	Mechanisms for environmental improvement	Agri-environmental policy configuration
Radical decoupling	The double dividend: price-induced extensification; reallocation of money into environmental schemes	Strictly decoupled environmental management schemes, payments calculated with reference to environmental outputs achieved
Moderate decoupling	Maintaining sufficient farming activity to ensure production of joint products; additional improvements engineered through top-up payments	Bottom tier hectarage payments and voluntary set aside, upper tier environmental payments offered on a discretionary basis
Weak decoupling	Voluntary enrolment in ELMS; application of cross compliance to producer subsidies and compensation	ELMS together with pragmatic use of conservation compliance

products required under what would be a reward-based system of payments. As recent experience with the EU's AEP has shown, there is a good nature conservation case for enrolling large numbers of farmers into basic entry level schemes in order to maximize policy reach and then using tiered payments to promote more ambitious land use changes as required. Arguably, this would become more difficult to achieve under a system which would seek to maximize the immediately measurable addition to environmental quality on every ECU spent.

Proposals were made some time ago for a more moderately decoupled policy which would allow agriculture departments to continue supporting large numbers of farmers in a less trade-distorting way than at present. Under a system of 'producer entitlement guarantees', for instance, farmers would be allocated an entitlement to an annual government subsidy calculated as a percentage of the output of their farms in the recent past, agricultural output above this level being sold on an unsupported market (Harvey, 1990). According to Harvey (1990, p. 212), 'if there are real social benefits from a more economically secure farming population than the free market would provide, then some annual payments may well be justified'. Allan Buckwell's (1996) Common Agricultural and Rural Policy of Europe (CARPE) envisages a gradual substitution of existing price support and compensation payments by widely available Environmental and Cultural Landscape payments and Rural Development Assistance. Under a three tiered agri-environmental policy, farmers would be required to meet basic environmental standards without compensation but would also be eligible to apply for lower tier payments available 'over large parts of the European territory'. Upper tier payments would be more selectively deployed to bring about more ambitious environmental management on farms. Transitional assistance would be available 'whilst farmers receive and digest the message that society will pay market prices for market products and reasonable prices for the non-market services they provide but no more than this' (Buckwell, 1996, p. 15). Another not dissimilar idea, informed by a stronger precautionary philosophy, is to offer to all farmers who will take it an area payment, again in the form of a permanent policy entitlement but subject to conservation compliance, and then to employ tiering to secure any additions to environmental quality that may be required (see Table 6.2). While such a system is some way from the fully decoupled arrangement envisaged by Jenkins and others, it could be argued that it is more in line with the broader socio-environmental priorities of member states and thus more conceivable within the context of a European Rural Policy of the sort canvassed at Cork. In the UK, such thinking is beginning to find expression in the idea of protecting and enhancing 'countryside character' but it has always been to the fore in Germany and France where today it translates into a requirement to retain a sufficiently broad base of government support to sustain a strong agricultural presence in rural areas. The debate about the defensibility of this strategy in WTO terms has yet to take place.

The liberalization of agricultural trade is now firmly on the international policy agenda. While movement in this direction may presently be rather uneven, it is the common goal of WTO members that agricultural support should progressively be scaled down and decoupled in the years ahead. This defines the background against which the future development of agri-environmental policy will need to take place. There is an argument that liberalization will be doubly beneficial for the rural environment, easing the pressure for intensification created by high price support and freeing up money for an expansion in green subsidies. This is an attractive idea because it suggests that improving agricultural competitiveness and boosting environmental protection are complementary policy goals. In reality, there are grounds for caution concerning this double dividend equation. In a US context, the much vaunted extensification effect may yield only marginal environmental benefits when set against the likely increased erosion hazard of having more land in production once the commodity programmes have been wound up. For the European countryside, the benefits of reduced pollution due to more extensive production need to be balanced against the likely loss in biodiversity and cultural landscape values due to a decline in countryside management and the economic marginalization of high natural value farming systems. While these side-effects could be dealt with through an expansion in agri-environmental programmes, it is unclear how willing or able policymakers will be to achieve the green recoupling of farm support that would seem to be required.

In the first sign of a serious divergence in their agri-environmental policy trajectories, the US and EU look set to respond in different ways to the challenges of liberalization. American policymakers, judging by pronouncements made since the passage of FAIR, are overwhelmingly concerned to preserve farmers' freedom to farm in order to exploit the export opportunities that are opening up for US agriculture in a post-GATT world. At the same time, budgetary constraints and federal rollback make it unlikely that there will be any further expansion in federally funded land retirement programmes. This leaves them with a rather polarized set of policy choices (Swanson, 1993). More Congressional and USDA control over the operation of surviving soil conservation programmes will mean an increasingly selective (and more radically decoupled) deployment of the CRP and its sister programmes like the WRP. Elsewhere, the emphasis will likely shift to the farm level, where individual farmers will be encouraged to find their own managerial and technological solutions to the environmental problems of intensively cropped land. The resulting policy vacuum here is likely to be filled by more state regulation and a trend towards the imposition of safe minimum standards on public health, food safety and environmental grounds. For European policymakers, on the other hand, the policy choice set is much less clear cut and, indeed, has still to be properly worked out. The challenge here will be to balance the need to meet international obligations under the WTO to decouple CAP support, with the desire to retain a broad base of support in order to preserve the social and environmental fabric of a managed countryside much more vulnerable to the restructuring effects of world market forces. Support for a redirection of funds into schemes that are exclusively environmental in purpose is likely to be tempered with a concern to develop a new European rural policy designed to protect the social structure of rural areas as well as the landscapes and habitats they contain. A second-best strategy could be emerging, involving a series of evolutions towards a more moderately decoupled policy compared with the US. While many traditional policy entitlements may well be preserved, they are likely to be made more subject to environmental obligations and quasi-regulatory codes of good agricultural practice designed to give Europe's citizens the countryside they desire and for which they are still apparently willing to pay.

7 The Environmental Reform of Farm Policy?

It would be too much to claim that the agri-environmental reforms of the past decade have brought the environment into the heart of farm policy. On any objective reckoning, government spending on agri-environmental programmes and schemes in the EU and US is still modest compared with the allocations being made for market support and, increasingly, compensation for reductions in market support. Judging by their actions, policymakers remain reluctant to substitute green support for conventional production aids to any significant degree, and there is a sense in which agri-environmental policies continue to be seen as an 'accompanying measure' to what is essentially a productivist system of agricultural support. While there has been a proliferation of green payment schemes in all the countries concerned, these are typically cross cut by price subsidies, production aids and compensation measures. The approach, it would seem, at least until very recently, has been to add new environmental measures without taking away the agricultural policies deemed to be a significant cause of the problem. Winters's (1987, p. 300) comment that in the agricultural policy field there is always a greater tendency towards 'doing something' than towards 'undoing something', with a resulting 'costly tangle of potentially inconsistent instruments' appears well confirmed here. Indeed, as this book has shown, many current agri-environmental programmes have themselves a double or even triple purpose, their green credentials being a cover for the pursuit of more traditional policy goals like the support of farmers' incomes and the control of overproduction.

This last feature should come as no surprise to those who hold that policy causes are linked to policy outcomes. It is clear that policymakers had mixed motives in agreeing to set up the first green payment schemes and that this has been reflected in their design and subsequent implementation. The mid-1980s were crisis times for the agricultural policy community as overproduction and

154

a downturn in world market prices put severe pressure on farm budgets. To this was added an environmental critique of agricultural policy which identified market support as one of the main driving forces behind environmental degradation and decline. The challenge from environmentalists was important and free standing because it appeared to question the legitimacy of the claim, long proclaimed in both the US and Europe, that farmers deserved support because of their role as producers of public goods, including environmental ones. So deep seated was this view that it barely needed to be articulated, let alone defined, by the policy community concerned. The environmental critique constituted a potentially serious attack on one of the key legitimizing assumptions of government support for agriculture – that farmers need to be financially secure if they are to carry out the stewardship role that has been assigned to them. In the final analysis, however, it is debatable whether the newly mobilized agri-environmentalists succeeded in breaching the defences of the agricultural establishment, at least during this first wave of policy change. As Bonnen and Browne (1989) point out, accommodation is always likely to be the first instinct of agricultural policymakers in industrial countries, with each new policy participant being given some influence over a narrow part of the whole. Established agricultural interest groups, for their part, sometimes concede some 'policy turf' to new claimants, but usually on their own terms. In the agri-environmental policy case, green reforms came to be seen by all parties as a 'magic bullet' (Vail *et al.*, 1994), capable of making a contribution to supply control and offering an additional income source to farmers as well as protecting the environment. This was a critical perception, allowing policy actors to unite around a common, if partial, programme of green reform. Its practical effect, however, was a series of reforms which have attempted to look both ways at once, having been designed to achieve implicit agricultural as well as explicit environmental policy goals.

The resulting policy contradiction can be seen most clearly in the US, where the Conservation Title of the 1985 Farm Act introduced large ideas like the Conservation Reserve and conservation compliance into agricultural policy to tackle a soil erosion problem that was widely agreed to be a federal government responsibility, and yet failed to ensure that conservation would be kept to the fore during their implementation. The Secretary of Agriculture was given extensive discretion over the CRP implementation decision that mattered most – which land to set aside – and anxious to maximize the supply control impact of the policy, and keen to appease the farm lobby by spreading payments, USDA proceeded to enrol land on a wide front. Conservation compliance almost suffered a similar fate, its effectiveness diluted by the SCS's decision to water down the soil conservation standards farmers had to fulfil in order to continue receiving federal government support. Here, and elsewhere, there is evidence of a classic implementation gap in policy which arose in part because of too close an identity of interest between the agencies responsible for operating the policy and its recipients, the farmers. (This may yet rebound on USDA, leading to its replacement by other agencies like the EPA which can combine a more disinterested policy stance with the institutional reach at present thought to be unique to the agricultural Extension Service.) The result of this gap was a much less rigorous prosecution of the Title than the conservation coalition which helped frame it would have liked and, arguably, the gradual subversion of its conservation rationale. There is a certain amount of what Keeler (1996) calls 'path dependence' in this story, given the way in which recent policy has followed tracks and embodied assumptions laid down in previous periods. The Conservation Title was supposed to be a radical departure for US soil conservation policy, yet it centred on the New Deal idea of paying farmers to take cropland out of production in order to address the very traditional problem of soil erosion on farms. The established policy mix of voluntary compliance, technical assistance and incentive payments remained more or less intact. Admittedly, the linked mechanisms of compliance, sodbuster and swampbuster were much more of a departure, to the extent that they represented a new form of quasi-regulation whose appearance could not have been predicted from previous experience. Ultimately, however, their impact has been eroded as farmer dependence on commodity support has declined. As Swanson (1993, p. 112), reviewing the immediate impact of the 1985 Conservation Title puts it, because its crucial targeting provisions were never fully observed, 'this legislation did not diverge from the traditional path of subordinating conservation to the political goal of income maintenance, in this case by reducing the number of acres under tillage'.

European policymakers began with a much cleaner sheet when they came to set up the EU's agri-environmental policy in 1992. They did however carry with them long-held, if more abstract, assumptions about the role of farmers as producers of public goods and were agreed on the need to ensure the continued agricultural occupancy of rural space in order to maintain its biodiversity, amenity interest and social stability. As in the US, the emergence of a consensus about the nature of the problem and how best to deal with it was an important part of the policy process. By recognizing farmers as potential producers of the countryside, policymakers were able to justify an essentially voluntaristic strategy of subsidizing them for the environmental services they provided. Farmers were to be paid to manage the countryside – a rather alien concept in American terms but one that had already been field tested in the UK, Denmark and The Netherlands – and to extensify production as a solution to, and (possibly more importantly) precaution against, agricultural pollution. The resulting measures, one plank in a larger raft of CAP reforms, was hailed by supporters as the culmination of the green reforms of the 1980s, not only in terms of the injection of 'new' money into the conservation of the European countryside and the prevention of pollution, but as part of the renegotiation of the social contract between farmers and the state that had last been defined under the productivist Treaty of Rome. The advent of an agri-environmental policy, it was claimed, paved the way for a pan-EU approach to the protection of the countryside and, taken together with the EU's Habitat Directive, had the potential to contribute significantly towards meeting the obligations of member states under the Biodiversity Convention. Meanwhile, the use of environmental contracts made possible by the new Regulation, meant that the EU was embarked on one of the largest experiments in the use of quasi-market incentives to influence and reward environmental behaviour and encourage innovation in the private production of public goods.

Against this, critics could still point to the fact that projected EC agri-environmental expenditure was dwarfed by the much larger sums allocated to farmers under the price support arm of the CAP and the compensation schemes set up under the same MacSharry reforms that had ushered in the Regulation itself. Indeed, compared with the other Accompanying Measures, agri-environmental expenditure did not dominate. In the implementation of the programme, doubts have been expressed about the extent to which environmental protection has been subsumed to the more traditional policy goal of maintaining farmers' incomes and there has been a growing sense that the national programmes are not all they seem. Countryside management and extensification have proved highly elastic concepts, with environmental outputs that are often hard to measure and difficult to judge, and this has been exploited by agriculture departments to establish measures that often do little more than subsidize existing good agricultural practice. In the other direction, where efforts have been made to bring about genuine additions to countryside management and environmental protection, administration and transaction costs have been high, raising questions about the extendibility of the approach to a wider community of farmers. There are also signs of a gap in attitude and approach between those (chiefly northern member states) choosing to operate an agri-environmental policy and those now obliged to do so under the Regulation. Certainly the wider deployment of agri-environmental schemes in southern member states, where institutional structures and delivery systems appear much less well developed, raises questions about the extent to which administrative control should be devolved to individual member states. The conventional wisdom is that a high degree of 'subsidiarity' is essential if measures are to be matched to the EU's extremely varied agri-environmental profiles and priorities. Nevertheless, a loss of central administrative control makes it more difficult for the Commission to insist on the improvements to policy design that may be necessary to ensure the policy is defensible at a later date in the context of the WTO (see below).

Fortunately, this is not the end of the story. Indeed it is probably only the beginning. There are a number of reasons to expect the further evolution of agri-environmental policy, albeit along increasingly divergent paths, in both the US and the EU. To begin with, policymakers have shown themselves willing to learn from past mistakes in order to improve the performance of the programmes already in place. As Hall (1993) points out, policy learning is an established part of the policymaking process and an essential component of an evolutionary (as against a revolutionary) model of policy change. It can be defined as a 'deliberate attempt to adjust the goals or techniques of policy in response to past experience and new information. Learning is indicated when policy changes as a result of such a process' (Hall, 1993, p. 278). A key motivation for such adaptation is the need to improve policy effectiveness in order to defend it against budgetary disciplines. This is particularly likely to apply to agri-environmental policy, funded as it is under expenditure headings that tend to be both more transparent and discretionary than the agricultural ones they replace. Policy learning is well advanced in the US, where early critical assessments of the CRP's environmental performance has already led to its virtual reinvention as a more strictly defined conservation policy tool. Improvements in the policy-relevant knowledge base about soil erosion, its geographical incidence and off-farm impacts were important factors here, but so too was the gaining perception that the policy would not be sustainable politically unless it delivered cost-effective improvements in environmental quality. This has been accompanied by a co-evolutionary shift in policy priorities, the need to prevent agricultural pollution and protect wildlife on farms beginning to eclipse the traditional farm-centred concern with soil erosion and its impact on productivity. With the selective recommissioning of the CRP under the 1996 FAIR Act, legislators have sent a signal that this and selected other federally funded programmes will continue, but subject to tighter performance standards than ever before. To adapt Zinn's (1994) analogy, if the 1985 Conservation Title was the first jolt in the agricultural policy landscape for 30 years, the result of a gradual build-up in pressure over a number of years, FAIR was something of an earthquake which only the most robust federal conservation programmes and structures will be able to survive.

Among the most important lessons of the 1980s is that the USDA and its agencies need to be prepared to 'micro-manage' federal programmes if they are serious about using them to deliver lasting and cost effective environmental improvements on farms. Effectively, this means being willing to draw on the best available scientific knowledge to target subsidies at regions and locations with the most acute environmental problems or potential, even if this subtracts from their role as income supports. Although it took some time to gain acceptance in policymaking circles, targeting is now an established practice, shifting an increasing proportion of soil conservation expenditure into the hands of farmers who promise to offer the best environmental return. The other principal lesson, however, is that while subsidies buy conservation, they rarely bring about lasting changes in farmer behaviour. This perception is growing now that the CRP contracts are beginning to expire and conservation compliance is set to decline as farm support is further scaled down. As was just seen in Chapter 6, it is likely to bring in its wake a greater readiness to use regulations to secure the permanent observance of safe minimum standards and to help force through lasting, technologically driven, changes to farming practice in a new era of environmental stewardship.

Budgetary disciplines will also be brought to bear on the EU's AEP in the years ahead, judging by the tone of an amending Regulation issued by the EC in 1996. This speaks of the need for further improvements in the design and delivery of schemes to ensure they subsidize activities which go beyond good agricultural practice (CEC, 1996b). In one sense, the EU's programme has been a policy experiment from the beginning, designed to field test the rather surprising idea that governments can engineer changes in farming practice in order to protect agricultural landscapes and maintain countryside character. The more gradual emergence of this policy, compared with the American Conservation Title 'big bang', has allowed the lead member states, at least, to edge towards the idea of green recoupling on a larger scale, developing delivery systems and sensitizing farmer opinion to the idea of being paid to 'produce countryside'. The UK especially, for reasons of political culture and institutional structure, has adopted a determinedly incrementalist approach, developing a portfolio of schemes to maximize the scope for policy learning and adopting small but important design improvements like discretionary targeting and tiered payment systems to achieve better environmental results. It is also here that a wider debate about greening the CAP has arguably been most advanced, and where interest in borrowing ideas from the US like conservation compliance most sustained. Elsewhere, agri-environmental policy has overlapped in a more obvious sense with existing conservation policies and legislation, payments to farmers in The Netherlands, Germany and Denmark being deployed to improve compliance with, or ease of implementation of, pollution control regulations and conservation standards. Meanwhile, the accession to the EU of countries like Sweden, with its own independent and rather sophisticated history of agri-environmental policy development, further enriches the policy mix, bringing new policy ideas into play. To adapt Nelson and Soete's (1988) observation, policy evolution from such a broad base offers much scope for engaging in experimentation and learning from feedback. The increasingly widespread introduction of tiered payments in member states is a case in point because it exploits the current situation of widespread but essentially shallow participation in schemes to reward farmers who subsequently agree to undertake more restrictive operations and agree to be brought deeper into schemes. European policymakers are learning that there is a fine line between setting up voluntary schemes that are restrictive enough to effect an improvement in something as complicated as countryside management, and yet also permissive enough to attract sufficient numbers of farmers and enough land to make a difference.

For those member states like The Netherlands, Germany and the UK with the longest history of involvement in agri-environmental policy, a further distinction will increasingly need to be drawn between a change in policy outcomes – setting up the schemes and enrolling farmers – and policy results: the short- and long-run environmental benefits that are actually produced on farms and in fields. As in the US, the durability of the environmental improvements that can be produced under voluntary contracts is already the subject of debate, much complicated here by the difficulty of deriving reliable indicators of what it is that has been added as a result of scheme participation. It is Whitby's (1996) view that an evolution in the public's appreciation of what agri-environmental policy is supposed to be about will exert its own pressure for a further tightening in the design and assessment of schemes. With the conservation interest on farms coming to be seen as a growing stock of environmental capital that has been paid for out of public funds, policymakers can be expected to take a more rigorous approach to evaluating the impact of programmes, defining ever more precisely the product to be delivered and locking farmers into schemes for the longer term. In the meantime, there is justifiable concern about the limited changes in farmer attitudes and behaviour current schemes appear to be bringing about. Few doubt that changing the culture of farming so that it is more sympathetic to the state of the environment should be one of the longer term objectives of agri-environmental policy, for as Colman *et al.* (1992, p. 69) suggest, 'policy measures which encourage positive attitudes to conservation will in the long term be more effective than those that do not, since a positive shift in attitudes will increase the output of conservation goods at any specified level of budgetary cost'. To this could be added that unless they exert such an influence, agri-environmental policy measures will inevitably be seen as temporary bribes, shallow in operation and transitory in impact. Their policy shelf-life will be very short indeed.

Further ahead, there are changes likely to the external policy environment which will force another, probably more accelerated, evolution of agri-environmental policy. A key catalyst is the international liberalization of agricultural trade under the aegis of a newly constituted WTO. This is a qualitatively different project from the domestic reform of farm policy to keep it within budgetary constraints that has framed the greening process up until now. It requires policymakers to simplify and withdraw support, not complicate and add to it. Moreover, it is driven by an ideological commitment to global free trade and requires convergence towards a common strategy for agricultural policy reform. As Weale (1992) remarks, internationally agreed rules can impose their own influence over actions and events, redefining the limits of the possible as well as imposing constraints. In this case, they take the form of WTO rules and disciplines governing how agricultural policy reform should proceed and at what pace. Specifically, these require policymakers to improve market access and decouple the subsidies farmers receive from their production decisions. With the decoupling process – probably the most fundamental change in the design of farm policy for 40 years – are likely to come new pressures to re-legitimize the basis of agricultural subsidies as the agricultural policy community, especially in Western Europe, finds itself put in the position of having to defend the idea that farmers should continue to receive money from the public purse. This prospect is already introducing a stronger dynamic into agri-environmental reform as the farm lobby seeks to renew its alliance with environmentalists to promote the long-term 'green recoupling' of support. There is a growing sense that the greening process is about to enter a new phase.

If it does, then the EU and US may find themselves following radically different paths so far as future agri-environmental policy development is concerned. The US, an instigator of the Uruguay Round and a keen advocate of further agricultural liberalization, has already raised the stakes of the next round of WTO talks by passing the FAIR Act into domestic law. This legislation, and the decoupling of farm policy it brings about, signals that the US intends to pursue a strong line so far as the definition of 'green box' subsidies and the paring down of agricultural support is concerned. In order to be credible in promoting this strategy internationally, US policymakers will be under pressure to further streamline federal conservation programmes and decouple them from agricultural policy objectives such as the support of farmers' incomes. Consistent with a less public, more managerial approach to agri-environmental issues, there will be a growing interest in the scope for engineering environmental improvements into farming systems. By allowing land to be retained in production, private stewardship, widely exercised, it is claimed, allows agri-environmental problems to be addressed without compromising the export potential of US agriculture. The challenge for EU policymakers, by contrast, will be to defend a countryside more vulnerable environmentally, socially and culturally, to the restructuring effects of world market forces. It may be neither politically feasible or environmentally desirable to move to a fully decoupled policy here, and there will be resistance on social as well as environmental grounds to the idea of a large-scale withdrawal of government support. The most likely evolution of EU policy could involve policymakers agreeing to attach progressively more restrictive conditions to any income and compensation payments farmers would continue to receive during a long transition to a more liberal market regime. These may eventually give way to a permanent system of 'green box' hectarage payments or some other entitlement that would be offered to farmers on a conditional basis under a reformed European Rural Policy. While such a desideratum still looks to be some way off, it is at least now under serious discussion in European policymaking circles and has become a reference point for environmental lobbying in Brussels. Such an arrangement has the political advantage that it would dovetail with what is perceived by many commentators (see House of Commons, 1997) to be growing public pressure for safe minimum standards in agriculture. Policymakers will be increasingly interested in defining the agri-environmental quality standards – for soil and water quality, habitat protection and landscape maintenance – which farmers must achieve in order to qualify for continued support. It would also be consistent with the view that public support to Europe's farmers is justifiable as part of a continuing social contract, albeit one that is now to be couched in terms of individual merit rather than collective desert. A further evolution could see these common standards acquiring legal status and being enforced on a polluter pays basis, suitably modified according to local conditions and the requirements of fair competition. By this stage, farmers would be required to meet further conditions or achieve additional improvements in order to qualify for support, and then to go beyond the standard in order to enter a second tier of higher payments, possibly funded from national exchequers on a discretionary basis. These latter could be offered in the form of an environmental contract modelled on the more successful agri-environmental schemes which currently exist. They would be offered for the maintenance and restoration of environmental capital as well as the provision of services.

What is clear at the conclusion of this book is that agri-environmental reform is far from complete and still has a long way to run. Policy logic and WTO rules, to say nothing of shifting public opinion, suggest that green payments will be one of the few politically sustainable forms of government support to agriculture in the years ahead and that an agri-environmental policy in some form is set to become a permanent part of the rural policy scene. Important questions remain about the correct balance between subsidization and regulation within this evolving framework, about taxpayers' willingness to fund government programmes which may not always deliver an immediate or measurable environmental benefit, and of the willingness and ability of farmers themselves to take on the role of environmental stewards that is being prepared for them. Particular uncertainty surrounds the process of agricultural liberalization to which most industrial countries are now committed, and the extent to which policymakers will be able to extract a conservation dividend from the withdrawal of conventional farm support. It would appear that agri-environmentalists are about to embark on the long game of farm policy reform, played at greater risk but for high stakes.

References

Abler, D. and Shortle, J. (1992) Environmental and farm policy linkages in the US and the EC. *European Review of Agricultural Economics* 19 (2), 197–217.

Adams, W. (1986) *Nature's Place: Conservation Sites and Countryside Change.* Allen & Unwin, London.

Adams, W., Bourn, N. and Hodge, I. (1992) Conservation in the wider countryside. SSSIs and wildlife habitat in eastern England. *Land Use Policy* 9 (4), 235–248.

AFT (1984) *Soil Conservation in America. What Do We Have to Lose?* American Farmland Trust, Washington, DC.

AFT (1996) Agricultural Policy Proposals for the 1995 Farm Bill. Unpublished, American Farmland Trust, Washington, DC.

Agra-Europe (1991) *Agriculture and the Environment: How Will the EC Resolve the Conflict?* Agra-Europe, Tunbridge Wells, Kent.

Alexander, R. (1989) *Offsite Sediment Benefits of the Conservation Reserve Program in the Southern United States.* University of Tennessee Agricultural Economics and Rural Sociology, Tennessee.

Anderson, K. (1992) Agricultural trade liberalisation and the environment: a global perspective. *The World Economy* 15 (1), 153–171.

Anderson, K. and Hyami, Y. (1986) *The Political Economy of Agricultural Protection.* Allen & Unwin, Sydney.

Anderson, K. and Strutt, A. (1996) On measuring the environmental impact of agricultural trade liberalisation. In: Bredahl, M., Ballenger, N., Dunmore, J. and Roe, T. (eds), *Agriculture, Trade and the Environment: Discovering and Measuring the Critical Linkages.* Westview Press, Boulder, Colorado, pp. 151–172.

Ayer, H. (1996) FAIR: Key Commodity, Trade and Conservation Provisions. Unpublished paper to Agricultural Economics Society Conference 'The US Farm Bill: Implications for Future CAP Reform', 26 November, London.

Babcock, B. (1996) Future applications of the CRP. Unpublished Statement to the US Senate Committee on Agriculture, Nutrition and Forestry, Washington, DC.

Baker, G., Rasmussen, W., Wiser, V. and Porter, M. (1963) *Century of Service: the First One Hundred Years of the USDA*. USDA, Washington, DC.

Baker, S., Milton, K. and Yearly, S. (eds) (1994) *Protecting the Periphery: Environmental Policy in Peripheral Regions of the European Union*. Frank Cass, Ilford.

Baldock, D. (1990) *Agriculture and Habitat Loss in Europe*. CAP Discussion Paper No. 3, WWF International, Gland, Switzerland.

Baldock, D. (1992) The polluter pays principle and its relevance to the European Community's agricultural policy. *Sociologia Ruralis* 32 (1), 49–65.

Baldock, D. (1994) The OECD experience of integrating agricultural and environmental policy. In: *Agriculture and the Environment in the Transition to a Market Economy*. OECD, Paris, pp. 43–67.

Baldock, D. and Beaufoy, G. (1992a) *Green or Mean? Assessing the Environmental Value of CAP Reform 'Accompanying Measures'*. Council for the Protection of Rural England, London.

Baldock, D. and Beaufoy, G. (1992b) *Plough On! An Environmental Appraisal of the Reformed CAP*. World Wide Fund for Nature, Godalming.

Baldock, D. and Bennett, G. (1991) *Agriculture and the Polluter Pays Principle in Six EC Countries*. Institute for European Environmental Policy, London.

Baldock, D. and Conder, D. (eds) (1985) *Can the CAP Fit the Environment?* Institute for European Environmental Policy/Council for the Protection of Rural England/World Wide Fund for Nature, London.

Baldock, D. and Lowe, P. (1996) The development of European agri-environment policy. In: Whitby, M. (ed.) *The European Environment and CAP Reform: Policies and Prospects for Conservation*. CAB International, Wallingford, pp. 8–25.

Baldock, D. and Mitchell, K. (1995) *Cross-Compliance within the Common Agricultural Policy: a Review of Options for Landscape and Nature Conservation*. Institute for European Environmental Policy, London.

Baldock, D., Cox G., Lowe, P. and Winter, M. (1990) Environmentally Sensitive Areas: incrementalism or reform? *Journal of Rural Studies* 6 (2), 143–162.

Baldock, D., Beaufoy, G., Bennett, G. and Clark, J. (1993) *Nature Conservation and New Directions in the EC Common Agricultural Policy*. Institute for European Environmental Policy, London.

Barr, C., Bunce, R., Clarke, R., Fuller, R., Furse, M., Gillespie, M., Groom, G., Hallman, C., Hornung, M., Howard, D. and Ness, M. (1993) *Countryside Survey 1990, Main Report*. Department of the Environment, London.

Batie, S. (1982) Policies, institutions and incentives for soil conservation. In: Halcrow, H., Heady, E. and Cotner, M. (eds), *Soil Conservation Policies, Institutions and Incentives*. Soil Conservation Society of America, Ankeny, pp. 25–40.

Batie, S. (1984) *Soil Erosion. Crisis in America's Croplands?* The Conservation Foundation, Washington, DC.

Batie, S. (1985) Soil conservation in the 1980s: a historical perspective. *Agricultural History* 59, 107–123.

Batie, S. (1986) Conservation cross compliance: an alternative perspective. Unpublished paper to the Annual Agricultural Outlook Conference, US Department of Agriculture, Washington, DC.

Batie, S. (1988) Agriculture as the problem: new agendas and new opportunities. *Southern Journal of Agricultural Economics* 20 (1), 1–12.

Batie, S. (1990) Agricultural policy and environmental goals: conflict or compatibility? *Journal of Economic Issues* 24 (2), 565–573.

Batie, S. (1994) Coordinating agricultural conservation. In: *When Conservation Reserve Program Contracts Expire: the Policy Options.* Soil and Water Conservation Society, Ankeny, pp. 127–129.

Baudry, J. and Laurent, C. (1993) Paysages ruraux activites agricoles. In: Courtet, C., Berlan-Darque, M. and Demarne, Y. (eds), *Agricultures et Societie: Pistes pour la Recherche.* Association Descartes, Paris.

Beaufoy, G., Baldock, D. and Clark, J. (1994) *The Nature of Farming. Low Intensity Farming Systems in Nine European Countries.* Institute for European Environmental Policy/World Wide Fund for Nature/Joint Nature Conservation Committee, London.

Benbrook, C. (1979) Integrating soil conservation and commodity programs: a policy proposal. *Journal of Soil and Water Conservation* 34 (4), 160–167.

Benbrook, C. (1980) An examination of the fledgling alliance of soil conservation and commodity price support programs. *North Central Journal of Agricultural Economics* 2 (1), 1–16.

Benedict, M. (1953) *Farm Policies of the United States, 1790–1950.* The Twentieth Century Fund, New York.

Bennett, H. (1931) The problem of soil erosion in the United States. *Annals of the Association of American Geographers* 21 (3), 147–170.

Bennett, G. (1984) The application of the LFA Directive in The Netherlands. In: *House of Lords Select Committee on the European Communities, 20th Report.* Session 1983–84, HMSO, London, pp. 171–180.

Berlan-Darque, M. and Teherani-Krönner, P. (1992) The ecologization of French agriculture. *Sociologia Ruralis* XXXII (1), 104–114.

Berner, A. (1994) Wildlife and federal cropland retirement programs. In: *When Conservation Reserve Program Contracts Expire: the Policy Options.* Soil and Water Conservation Society, Ankeny, pp. 127–129.

BirdLife International (1996) *Nature Conservation Benefits of Plans under the Agri-Environment Regulation (EEC 2078/92).* BirdLife International, Sandy.

Bjerke, K. (1991) An overview of the agricultural resources conservation program. In: Joyce, L.A., Mitchell, J.E. and Skold, M.D. (eds), *The Conservation Reserve – Yesterday, Today and Tomorrow.* USDA Forest Service General Technical Report RM-203, Washington, DC, pp. 7–26.

Bodiguel, M. and Buller, H. (1989) Agricultural pollution and the environment in France. *Tijdshcrift voor Sociaal Wetenschappelijke Onderzoek van de Landbouwe* 3, 217–239.

Boisson, J-M. and Buller, H. (1996) France. In: Whitby, M. (ed.), *The European Environment and CAP Reform: Policies and Prospects for Conservation.* CAB International, Wallingford, pp. 105–130.

Bonnen, J. and Browne, W. (1989) Why is agricultural policy so difficult to reform? In: Kramer, C.S. (ed.), *The Political Economy of U.S. Agriculture: Challenges for the 1990s.* Resources for the Future, Washington, DC, pp. 7–33.

Bowers, J. (1995) Sustainability, agriculture, and agricultural policy. *Environment and Planning A* 27 (8), 1231–1243.

Bowers, J. and Cheshire, P. (1983) *Agriculture, the Countryside and Land Use: an Economic Critique.* Methuen, London.

Bowler, I. (1985) *Agriculture Under the Common Agricultural Policy*. Manchester University Press, Manchester.

Brandow, G. (1977) Policy for commercial agriculture, 1945–71. In: Martin, L. (ed.), *A Survey of Agricultural Economics Literature*, Vol. I, *Traditional Fields of Agricultural Economics, 1940s to 1970s*. University of Minnesota Press, Minneapolis, Minnesota, pp. 209–292.

Britton, D. (1977) Some explorations in the analysis of long-term changes in the structure of agriculture. *Journal of Agricultural Economics*, 28, 697–702.

Brooks, J. (1996) Agricultural policies in OECD countries: what can we learn from political economy models? *Journal of Agricultural Economics* 47 (3), 366–389.

Brooks, J. and Carter, C. (1994) *The Political Economy of US Agriculture*. Research Report 94.8, ABARE, Canberra.

Brotherton, I. (1991) What limits participation in ESAs? *Journal of Environmental Management* 32 (3), 241–249.

Brouwer, F. and van Berkum, S. (1996) CAP and the environment in the European Union. Unpublished paper presented to conference, European Agriculture at the Crossroads, May 1996, University of Crete, Rethimno.

Browne, W. (1988) *Private Interests, Public Policy, and American Agriculture*. University Press of Kansas, Lawrence, Kansas.

Brubaker, S. and Castle, E. (1982) Alternative policies and strategies to achieve soil conservation. In: Halcrow, H., Heady, E. and Cotner, M. (eds), *Soil Conservation Policies, Institutions, and Incentives*. Soil Conservation Society of America, Ankeny, pp. 302–314.

Bruckmeier, K. and Teherani-Krönner, P. (1992) Farmers and environmental regulation: experiences in the Federal Republic of Germany. *Sociologia Ruralis* 22 (1), 66–81.

Buckwell, A. (1989) Economic signals, farmers' response and environmental change. *Journal of Rural Studies* 5 (2), 149–160.

Buckwell, A. (1996) Towards a common agricultural and rural policy for Europe. Unpublished Winegarten Memorial Lecture, 10 December, Wye, Kent.

Buller, H. (1992) Agricultural change and the environment in western Europe. In: Hoggart, K. (ed.), *Agricultural Change, Environment and Economy. Essays in Honour of W.B. Morgan*. Mansell Publishing, London, pp. 68–88.

Buller, H. (1996) Towards sustainable water management: catchment planning in France and Britain. *Land Use Policy* 13 (4), 289–302.

Burch, F., Green, B., Mitchely, M. and Potter, C. (1997) Possible options for the better integration of environmental concerns in support for arable crops. Unpublished research report to the European Commission, Wye College, University of London.

Burt, T., Heathwaite, A. and Trudgill, S. (1993) *Nitrate: Processes, Patterns and Management*. John Wiley & Sons, Chichester.

Buttel, F. (1989) The US farm crisis and the restructuring of American agriculture: domestic and international dimensions. In: Goodman, D. and Redclift, M. (eds), *The International Farm Crisis*. Macmillan, London, pp. 46–83.

Buttel, F. (1992) Environmentalism: origins, processes, and implications for rural social change. *Rural Sociology* 57 (1), 1–27.

Buttel, F. and Swanson, L. (1986) Soil and water conservation: a farm structural and public policy context. In: Lovejoy, S. and Napier, T. (eds), *Conserving Soil: Insights from Socioeconomic Research*. Soil Conservation Society of America, Ankeny, pp. 26–39.

CC (1974) *New Opportunities for the Countryside.* Countryside Commission, Manchester.

CC (1984) Memorandum to the House of Lords Select Committee on European Communities. *Agriculture and the Environment.* 20th Report, Session 1983–84. HMSO, London, pp. 162–188.

CC (1989) *Incentives for a New Direction for Farming.* Countryside Commission, Manchester.

CC (1993) *Paying for a Beautiful Countryside: Securing Environmental Benefits and Value for Money from Incentive Schemes.* Countryside Commission, Manchester.

CC (1997) Memorandum by the Countryside Commission. In: House of Commons Select Committee on Agriculture, *Environmentally Sensitive Areas and Other Schemes under the Agri-Environment Regulation.* Vol. II, *Minutes of Evidence,* Second Report. Session 1996–97. HMSO, London, pp. 108–117.

CEC (1969) *Memorandum on the Reform of Agriculture in the EEC.* Supplement to Bulletin No. 1. Office for Official Publications of the European Communities, Luxembourg.

CEC (1983) *Adjustment of the Community Agricultural Policy.* Bulletin of the European Communities, 4/83. Office for Official Publications of the European Communities, Luxembourg.

CEC (1985a) *Perspectives for the Common Agricultural Policy.* Office for Official Publications of the European Communities, Luxembourg.

CEC (1985b) Regulation on improving the efficiency of agricultural structures. *Official Journal,* 28, 1–18. Office for Official Publications of the European Communities, Luxembourg.

CEC (1987) *Agriculture and Environment: Management Agreements in Four Countries of the European Communities.* Office for Official Publications of the European Communities, Luxembourg.

CEC (1988) *Environment and Agriculture, COM 88(338).* Office for Official Publications of the European Communities, Luxembourg.

CEC (1990) *Agriculture and the Environment.* Press release, 25 July 1990. Office for Official Publications of the European Communities, Luxembourg.

CEC (1991a) *The Development and Future of the CAP, COM 91(100).* Office for Official Publications of the European Communities, Luxembourg.

CEC (1991b) *The Agricultural Situation in the Community, 1990 Report.* Office for Official Publications of the European Communities, Luxembourg.

CEC (1992) *Council Regulation EEC 2078/92 on the Introduction and Maintenance of Agricultural Production Methods Compatible with the Requirements of the Preservation of the Environment and the Management of the Countryside.* Office for Official Publications of the European Communities, Luxembourg.

CEC (1995) *Agricultural Situation in the Community, 1994 Report.* Office for Official Publications of the European Communities, Luxembourg.

CEC (1996a) *The Cork Declaration: a Living Countryside.* European Conference on Rural Development, November 1996, Cork. Office for Official Publications of the European Communities, Luxembourg.

CEC (1996b) *Council Regulation 746/96 Laying Down Detailed Rules for the Application of Council Regulation 2078/92.* Office for Official Publications of the European Communities, Luxembourg.

CEC (1996c) *The Agricultural Situation in the Community,* 1995 Report. Office for Official Publications of the European Communities, Luxembourg.

Cheshire, P. (1985) The environmental implications of European agricultural support policies. In: Baldock, D. and Conder, D. (eds), *Can the CAP Fit the Environment?* Institute for European Environmental Policy/Council for the Protection of Rural England/World Wide Fund for Nature, London, pp. 9–18.

Cheshire, P. and Bowers, J. (1969) Farming, conservation and amenity. *New Scientist* 3 April, 13–15.

Christenson, R. (1959) *The Brannan Plan.* University of Michigan Press, Ann Arbor.

CLA (1994) *Focus on the CAP: a Discussion Paper.* Country Landowner's Association, London.

Clark, E., Haverkamp, J. and Chapman, W. (1985) *Eroding Soils: the Off-Farm Impacts.* The Conservation Foundation, Washington, DC.

Clark, J., Jones, A., Potter, C. and Lobley, M. (1997) Reconceptualising the evolution of the European Union's agri-environmental policy: a discourse approach. *Environment and Planning A* 7, 1869–1885.

Clark, R. and Johnson, J. (1990) Implementing the conservation title of the 1985 Food Security Act: conservation or politics? In: Napier, T. (ed.), *Implementing the Conservation Title of the Food Security Act of 1985.* Soil and Water Conservation Society, Ankeny, pp. 26–48.

Cochrane, W. (1958) *Farm Prices: Myth and Reality.* University of Minnesota Press, Minneapolis, Minnesota.

Cochrane, W. (1980) Some nonconformist thoughts on welfare economics and commodity stabilization policy. *American Journal of Agricultural Economics* 62 (3), 508–511.

Cochrane, W. and Runge, C. (1992) *Reforming Farm Policy: Toward a National Agenda.* Iowa State University Press, Ames, Iowa.

Colman, D. (1983) The free trade alternative. In: Korbey, A. (ed.), *Agriculture: the Triumph and the Shame, an Independent Assessment.* Centre for Agricultural Strategy, University of Reading, pp. 43–57.

Colman, D. and Traill, W. (1984) Economic pressures on the environment. In: Korbey, A. (ed.), *Investing in Rural Harmony: a Critique.* Centre for Agricultural Strategy, University of Reading, pp. 30–41.

Colman, D., Crabtree, J., Froud, J. and O'Carroll, L. (1992) *Comparative Effectiveness of Conservation Mechanisms.* Department of Agricultural Economics, University of Manchester, Manchester.

Conrad, J. (1988) Nitrate debate and nitrate pollution in FR Germany. *Land Use Policy,* 5 (2), 202–212.

Conrad, J. (1990a) *Nitrate Pollution and Agriculture.* Avebury Studies in Green Research, Aldershot.

Conrad, J. (1990b) *Do Public Policy and Regulation Still Matter for Environmental Protection in Agriculture?* EUI Working Paper EPU No. 90/6. European University Institute, Florence.

Conservation Foundation (1988) *Protecting America's Wetlands: an Action Agenda.* The Conservation Foundation, Washington, DC.

Cook, K. (1983) Soil conservation: PIK in a poke. *Journal of Soil and Water Conservation* 38 (6), 475–476.

Cook, K. (1989) The environmental era of US agricultural policy. *Journal of Soil and Water Conservation* 44 (5), 362–366.

Cook, K. (1994) The CRP's niche in agricultural conservation and environmental agendas – three perspectives. In: *When Conservation Reserve Program Contracts*

Expire: the Policy Options. Soil and Water Conservation Society, Ankeny, pp. 63–65.

Coppock, J. (1963) *North Atlantic Policy – the Agricultural Gap*. Twentieth Century Fund, New York.

Cox, G., Lowe, P. and Winter, M. (1985a) Changing directions in agricultural policy: corporatist arrangements in production and conservationist policies. *Sociologia Ruralis* 25 (2), 130–153.

Cox, G., Lowe, P. and Winter, M. (1985b) Land use conflict after the Wildlife and Countryside Act 1981: the role of the Farming and Wildlife Advisory Group. *Journal of Rural Studies* 1 (2), 173–183.

CPRE (1986) *Rearranging CAP Reform: Deck Chairs on the Titanic?* Council for the Protection of Rural England, London.

CPRE (1990) *Paradise Protection: How the European Community Should Protect the Countryside*. Council for the Protection of Rural England, London.

CPRE (1997) Memorandum to the House of Commons Select Committee on Agriculture, *Environmentally Sensitive Areas and Other Schemes Under the Agri-Environment Regulation*, Vol. II, *Minutes of Evidence*, Second Report, Session 1996–97. HMSO, London, pp. 82–94.

Crabtree, R. (1992) A more environmental CAP? Unpublished paper to Agricultural Economics Society Conference, 'Reform of the CAP in Relation to GATT', 10 December, London.

Creason, J. and Runge, C. (1990) *Agricultural Competitiveness and Environmental Quality: What Mix of Policies Will Accomplish Both Goals?* Centre for International Food and Agricultural Policy, University of Minnesota, St Paul, Minnesota.

Crosson, P. (1991) Cropland and soils: past performance and policy challenges. In: Frederick, K. and Sedjo, R. (eds), *America's Renewable Resources: Historical Trends and Current Challenges*. Resources for the Future, Washington, DC, pp. 173–203.

Crosson, P. (1996) Impact of environmental policies on the competitiveness of U.S. agriculture. Unpublished paper to conference, European Agriculture at the Crossroads, May 1996, University of Crete, Rethimno.

Crosson, P. and Stout, A. (1983) *Productivity Effects of Cropland Erosion in the United States*. Resources for the Future, Washington, DC.

Crutchfield, S., Hanson, L. and Ribaudo, M. (1993) *Agriculture and Water Quality Conflicts: Economic Dimensions of the Problem*. Agriculture Information Bulletin 676, USDA, Washington, DC.

Dahlberg, K. (ed.) (1986) *New Directions for Agriculture and Agricultural Research: Neglected Dimensions and Emerging Alternatives*. Rowman and Allenheld, Ottowa.

Daniels, T. (1988) America's Conservation Reserve Program: rural planning or just another subsidy? *Journal of Rural Studies* 4 (4), 405–411.

Davidson, J. (1977) *Conservation and Agriculture*. John Wiley & Sons, Chichester.

de Putter, J. (1995) *The Greening of Europe's Agricultural Policy: the 'Agri-Environmental Regulation' of the MacSharry Reform*. Ministry of Agriculture, Nature Management and Fisheries/Agricultural Economics Research Institute, The Hague.

Deenihan, J. (1996) Unpublished Opening Address to European Conference on Rural Development, 7 November, Cork.

Delorme, H. (1987) An outline of French views on land conversion programmes. In: Baldock, D. and Conder, D. (eds), *Removing Land from Agriculture, the Implications for Farming and the Environment*. Council for the Protection of Rural England/

Institute for European Environmental Policy/World Wildlife Fund (UK), London, pp. 40–42.

Delpeuch, B. (1994) Ireland's agri-environmental programme in the European context. In: Maloney, M. (ed.), *Agriculture and the Environment.* Royal Dublin Society, Dublin, pp. 37–42.

Díaz, M., Campos, P. and Pulido, F. (1996) The Spanish dehesas: a diversity in land use and wildlife. In: Pain, D. and Pienkowski, M. (eds), *Farming and Birds in Europe.* Academic Press, London, pp. 178–209.

Dickason, C. and Magleby, R. (1993) *Erosion Reduction Benefits for US Soil Conservation.* Economic Research Service, USDA, Washington, DC.

Dicks, M. (1987) More benefits with fewer acres please! *Journal of Soil and Water Conservation* 42 (3), 170–173.

Dicks, M. (1994) Costs and benefits of the CRP. In: *When Conservation Reserve Program Contracts Expire: the Policy Options.* Soil and Water Conservation Society, Ankeny, pp. 39–45.

Dicks, M. and Grano, A. (1988) Conservation policy insights for the future. *Journal of Soil and Water Conservation* 43 (3), 148–151.

Dicks, M. and Vertrees, J. (1987) Improving the pay-off from the Conservation Reserve Program. In: Halbach, D., Runge, F. and Larson, W. (eds), *Making Soil and Water Conservation Work: Scientific and Policy Perspectives.* Soil Conservation Society of America, Ankeny, pp. 109–120.

Dicks, M., Reichelderfer, K. and Boggess, W. (1987) *Implementing the Conservation Reserve Program.* Economic Research Service Staff Report 129, USDA, Washington, DC.

Dixon, J. and Taylor, J. (1990) *Agriculture and Environment: Towards Integration.* Royal Society for the Protection of Birds, Sandy.

Dower, J. (1945) *Conservation of Nature in England and Wales: Report of the Wild Life Conservation Special Committee.* HMSO, London.

Dubgaard, A. (1993) Agriculture and the environment in the European Community. In: Williamson, C. (ed.), *Agriculture, the Environment and Trade – Conflict or Cooperation?* International Policy Council on Agriculture and Trade, Washington, DC, pp. 68–81.

Duchêne, F., Szczepanik, E. and Legg, W. (1985) *New Limits on European Agriculture: Politics and the Common Agricultural Policy.* Croom Helm, London.

Ekin, P., Folke, C. and Constanza, R. (1994) Trade, environment and development: the issues in perspective. *Ecological Economics* 9, 1–12.

Elliot, A. (1984) The view from Agriculture House. In: Korbey, A. (ed.), *Investing in Rural Harmony: a Critique.* Centre for Agricultural Strategy, University of Reading, Reading, pp. 24–29.

English Nature (1996) *Annual Report.* English Nature, Peterborough.

English Nature (1997) Memorandum to House of Commons Select Committee on Agriculture, *Environmentally Sensitive Areas and Other Schemes under the Agri-Environment Regulation,* Vol. II, *Minutes of Evidence,* Second Report, Session 1996–97. HMSO, London, pp. 121–144.

Ervin, D. (1985) *Conservation in the 1985 Farm Bill: Another Perspective.* Agricultural Economics Working Paper, University of Missouri–Colombia, Colombia, Missouri.

Ervin, D. (1988) Set Aside programmes: using US experience to evaluate UK proposals. *Journal of Rural Studies* 4 (3), 181–191.

Ervin, D. (1990) Economic efficiency effects and Government finance impacts of the conservation title: policy and program implications. In: Halcrow, H., Heady, E. and Cotner, M. (eds), *Soil Conservation Policies, Institutions, and Incentives*. Soil Conservation Society of America, Ankeny, pp. 67–80.

Ervin, D. (1993) Conservation policy futures: an overview. *Journal of Soil and Water Conservation* 48 (4), 421–424.

Ervin, D. and Dicks, M. (1988) Cropland diversion for conservation and environmental improvement: an economic welfare analysis. *Land Economics* 64 (3), 256–267.

Ervin, D. and Graffy, E. (1996) Leaner environmental policies for agriculture. *Choices*, Fourth Quarter 1996, 27–33.

Ervin, D. and Keller, V. (1996) Key questions. In: Bredahl, M., Ballenger, N., Dunmore, J. and Roe, T. (eds), *Agriculture, Trade and the Environment: Discovering and Measuring the Critical Linkages*. Westview Press, Boulder, Colorado, pp. 281–294.

Ervin, D., Algozin, K., Carey, M., Doering, O., Terichs, S., Heimlich, R., Hrubovcak, J., Konyar, K., McCormack, I., Osborn, C., Ribaudo, M. and Shoemaker, R. (1991) *Conservation and Environmental Issues in Agriculture*. Economic Research Service, USDA, Washington, DC.

Esseks, J. and Kraft, S. (1986) Landowner views of obstacles to wider participation in the Conservation Reserve Program. *Journal of Soil and Water Conservation* 41 (6), 410–414.

Ewins, A. and Roberts, R. (1992) The Countryside Premium Scheme for set-aside land. In: Clark, J. (ed.), *Set-Aside*. BCPC Monograph No. 50, British Crop Protection Council, Farnham, pp. 229–234.

Extzarreta, M. and Viladomiu, L. (1989) The restructuring of Spanish agriculture and Spain's accession to the EEC. In: Goodman, D. and Redclift, M. (eds), *The International Farm Crisis*. Macmillan, London, pp. 156–182.

Feist, M. (1978) *A Study of Management Agreements*. Countryside Commission, Cheltenham.

Felton, M. (1993) Achieving nature conservation objectives: problems and opportunities with economics. *Journal of Environmental Planning and Management* 36 (1), 23–31.

Felton, M. and Marsden, J. (1990) *Heather Regeneration in England and Wales: A Feasibility Study for the Department of the Environment*. NCC, Peterborough.

Fennell, R. (1979) *The Common Agricultural Policy and the European Community*. Oxford University Press, Oxford.

Fennell, R. (1985) A reconsideration of the objectives of the Common Agricultural Policy. *Journal of Common Market Studies* 23 (3), 257–276.

Fennell, R. (1987) Reform of the CAP: shadow or substance? *Journal of Common Market Studies* 26 (1), 61–77.

Finegold, K. (1982) From agrarianism to adjustment: the political origins of New Deal agricultural policy. *Politics and Society* 11 (1), 1–27.

Fischler, F. (1996) Europe and its rural areas in the year 2000: integrated rural development as a challenge for policy making. Unpublished Opening Speech to European Conference on Rural Development, 7 November, Cork.

FoE (1983) *Proposals for a Natural Heritage Bill*. Friends of the Earth, London.

FoE (1992) *Environmentally Sensitive Areas: Assessment and Recommendations*. Friends of the Earth, London.

Fottorino, E. (1990) L'agriculture moderne en accusation. *Le Monde*, 6 January.

Fraser, I. (1996) Quasi-markets and provision of nature conservation in agri-environmental policy. *European Environment* 6 (3), 95–101.

Freshwater, D. (1989) The political economy of farm credit reform: the Agricultural Credit Act of 1987. In: Kramer, C. (ed.), *The Political Economy of U.S. Agriculture: Challenges for the 1990s.* Resources for the Future, Washington, DC, pp. 105–139.

Froud, J. (1994) Upland moorland with complex property rights. The case of the North Peak. In: Whitby, M. (ed.), *Incentives for Countryside Management: the Case of Environmentally Sensitive Areas.* CAB International, Wallingford, pp. 81–104.

GAO (1977) *To Protect Tomorrow's Food Supply – Soil Conservation Needs Priority Attention.* Report to Congress by US Comptroller General, General Accounting Office, Washington, DC.

GAO (1989) *Conservation Reserve Could be Less Costly and More Effective.* Report to Congress by US Comptroller General, General Accounting Office, Washington, DC.

GAO (1990) *Farm Programmes: Conservation Compliance Provisions Could be More Effective.* Report to Congress by US Comptroller General, General Accounting Office, Washington, DC.

Gardner, B. (1996) *European Agriculture: Policies, Production and Trade.* Routledge, London.

Garrido, F. and Monyano, E. (1996) Spain. In: Whitby, M. (ed.), *The European Environment and CAP Reform: Policies and Prospects for Conservation.* CAB International, Wallingford, pp. 86–104.

Glasbergen, P. (1992) Agro-environmental policy: trapped in an iron law? A comparative analysis of agricultural pollution control in the Netherlands, the United Kingdom and France. *Sociologia Ruralis* 22 (1), 30–48.

Gould Consultants (1985) *Wildlife and Countryside Act 1981. Financial Guidelines for Management Agreements.* Department of the Environment, London.

Gould Consultants (1986) *Changes in Land Use in England, Scotland and Wales, 1985–2000.* Laurence Gould Consultants, London.

Grant, W. (1995) The limits of CAP reform and the option of renationalization. *Journal of European Public Policy* 2 (1), 1–18.

Green, B. (1985) *Countryside Conservation,* 2nd edn. George Allen & Unwin, Hemel Hempstead.

Green, B. (1986) Agriculture and the environment: a review of major issues in the UK. *Land Use Policy* 3 (3), 193–204.

Griffin, R. and Stoll, J. (1984) Evolutionary processes in soil conservation policy. *Land Economics* 60 (1), 29–39.

Gummer, J. (1991) Quoted in: Harvey, J. and Wilson, R. Gummer's last stand. *Farmers Weekly,* 15 February 1991.

Haigh, N. and Grove-White, R. (1985) Introduction and observations. In: Baldock, D. and Conder, D. (eds), *Can the CAP Fit the Environment? Proceedings of a Conference.* Council for the Protection of Rural England, Institute for European Environmental Policy, World Wide Fund for Nature, London, pp. 7–18.

Hall, P. (1993) Policy paradigms, policy learning and the state. *Comparative Politics,* April 1993, 275–294.

Harold, C. (1992) Would taxation or trade liberalisation reduce pollution from agricutural fertilizers? Unpublished MSc thesis, University of Minnesota, St Paul, Minnesota.

Harold, C. and Runge, C. (1993) GATT and the environment: policy research needs. *American Journal of Agricultural Economics* 75, 789–793.

Harrington, W. (1991) Wildlife: severe decline and partial recovery. In: Frederick, K. and Sedjo, R. (eds), *America's Renewable Resources: Historical Trends and Current Challenges.* Resources for the Future, Washington, DC, pp. 205–246.

Harrison, G., Rutherford, T. and Wooton, I. (1995) Liberalising agriculture in the European Union. *Journal of Policy Modelling* 17 (3), 223–255.

Harvey, D. (1990) *The CAP and Green Agriculture.* Green Paper No. 3, Institute for Public Policy Research, London.

Harvey, D. (1995) EU cereals policy: an evolutionary perspective. *Australian Journal of Agricultural Economics* 39 (3), 193–217.

Harvey, D. (1996) The US Farm Bill: fair or foul? Unpublished paper to Agricultural Economics Society conference, 'The American Farm Bill: Implications for CAP Reform', 26 November, London.

Harvey, D. and Thomson, K. (1985) Costs, benefits and the future of the Common Market agricultural policy. *Journal of Common Market Studies* 24 (1), 1–20.

Harvey, D. and Whitby, M. (1988) Issues and policies. In: Whitby, M. and Ollerenshaw, J. (eds), *Land Use and the European Environment.* Belhaven, London, pp. 143–177.

Hawke, N., Robinson, A. and Kovalena, N. (1993) Set-aside: its legal framework and environmental protection. *Environmental Law and Management* 7, 153–158.

Heady, E. (1984) The setting for agricultural production and resource use. In: English, B. (ed.), *Future Agricultural Technology and Resource Conservation.* Iowa University Press, Ames, Iowa, pp. 8–30.

Heady, E. and Allen, C. (1951) *Returns from Capital Required for Soil Conservation Farming Systems.* Research Bulletin No. 381, College of Agriculture, Iowa State University, Ames, Iowa.

Heimlich, R. and Langner, L. (1986) *Swampbusting: Wetland Conversion and Farm Programs.* Economic Research Service, USDA, Washington, DC.

Heimlich, R. and Osborn, C. (1993) After the Conservation Reserve Program: macroeconomics and post-contract program design. In: *Proceedings of the Great Plains Agricultural Council.* Great Plains Agricultural Council, Rapid City.

Heimlich, R. and Osborn, C. (1994) Buying more environmental protection with limited dollars. In: *When Conservation Reserve Program Contracts Expire: The Policy Options.* Soil and Water Conservation Society, Ankeny, pp. 83–97.

Helms, D. (1990) New authorities and new roles: SCS and the 1985 Farm Bill. In: Napier, T. (ed.), *Implementing the Conservation Title of the Food Security Act of 1985.* Soil and Water Conservation Society, Ankeny, pp. 11–25.

Henderson, N. (1992) Wilderness and the nature conservation ideal: Britain, Canada, and the United States contrasted. *Ambio,* 21 (6), 394–399.

Hertel, T. (1990) Ten truths about supply control. In: Allen, K. (ed.), *Agricultural Policies in a New Decade.* Resources for the Future, Washington, DC, pp. 153–169.

Hill, M., Aaronovitch, S. and Baldock, D. (1989) Non-decision making in pollution control in Britain: nitrate pollution, the EEC Drinking Water Directive and agriculture. *Policy and Politics,* 17 (3), 227–240.

Hill, P., Green, B. and Edwards, A. (1992) *The Cost of Care: the Costs and Benefits of Environmentally Friendly Farming Practices.* Royal Institution of Chartered Surveyors, London.

Hoban, T. and Clifford, W. (1994) Public attitudes about agricultural water pollution. In: Swanson, L. and Clearfield, F. (eds), *Agricultural Policy and the*

Environment: Iron Fist or Open Hand? Soil and Water Conservation Society, Ankeny, pp. 151–170.

Hodge, I. (1991) Incentive policies and the rural environment. *Journal of Rural Studies* 7 (4), 373–384.

Hodge, I. (1992) Supply control and the environment: the case for separate policies. *Farm Management* 8 (2), 65–72.

Hodge, I. (1996) On penguins on icebergs: the Rural White Paper and the assumptions of rural policy. *Journal of Rural Studies* 12 (4), 331–337.

Hoggart, K., Buller, H. and Black, R. (1995) *Rural Europe: Identity and Change.* Edward Arnold, London.

Höll, A. and von Meyer, H. (1996) Germany. In: Whitby, M. (ed.), *The European Environment and CAP Reform: Policies and Prospects for Conservation.* CAB Internationàl, Wallingford, pp. 70–85.

House of Commons (1985) *Operation and Effectiveness of Part II of the Wildlife and Countryside Act,* Vol. I, Select Committee on Environment, First Report, Session 1984–85. HMSO, London.

House of Commons (1997) *Environmentally Sensitive Areas and Other Schemes under the Agri-Environment Regulation.* Vol. I, *Report and Proceedings.* Select Committee on Agriculture Second Report, Session 1996–97. HMSO, London.

House of Lords (1984a) *Agriculture and the Environment.* Select Committee on the European Communities, 20th Report, Session 1983–84. HMSO, London.

House of Lords (1984b) *Agricultural and Environmental Research.* Select Committee on Science and Technology, 4th Report, Session 1983–84. HMSO, London.

House of Lords (1988) *Set-Aside of Agricultural Land.* Select Committee on the European Communities, 10th Report, Session 1987–88. HMSO, London.

House of Lords (1989) *Nitrate in Water.* Select Committee on the European Communities, 16th Report, Session 1988–89. HMSO, London.

House of Lords (1991) *Development and Future of the Common Agricultural Policy.* Select Committee on the European Communities, 16th Report, Session 1990–91. HMSO, London.

House of Lords (1994) *The Implications for Agriculture of the Europe Agreements.* Select Committee on the European Communities, 7th Report, Session 1993–94. HMSO, London.

Howarth, R. (1985) *Farming for Farmers? A Critique of Agricultural Support Policy.* Institute of Economic Affairs, London.

Huntings Surveys (1986) *Monitoring Landscape Change.* Huntings Surveys, Borehamwood.

Ingersent, K., Rayner, A. and Hine, R. (eds) (1994) *Agriculture in the Uruguay Round.* Macmillan, London.

Jenkins, T. (1990) *Future Harvest: the Economics of Farming and the Environment.* Council for the Protection of Rural England, London.

Johnson, R., Ekstrand, E., McKean, J.R. and John, K. (1994) The economics of wildlife and the CRP. In: *When Conservation Reserve Program Contracts Expire: the Policy Options.* Soil and Water Conservation Society, Ankeny, pp. 45–51.

Jordan, G., Maloney, W. and McLaughlin, A. (1994) Characterizing agricultural policy-making. *Public Administration* 72 (4), 502–526.

Josling, T. (1994) The reformed CAP and the industrial world. *European Review of Agricultural Economics* 21 (3/4), 513–527.

Judt, T. (1996) *A Grand Illusion? An Essay on Europe.* Penguin, London.

Just, R., Lichtenberg, E. and Zilberman, O. (1991) Effects of feed grain and wheat programs on irrigation and groundwater depletion in Nebraska. In: Just, R. and Bockenstaal, W. (eds), *Commodity and Resource Policies in Agricultural Systems*. Springer, New York, pp. 215–232.

Kahn, J. and Kemp, W. (1985) Economic losses associated with the degradation of an ecosystem: the case of submerged vegetation in Chesapeake Bay. *Journal of Environmental Economics and Management* 12, 246–263.

Keeler, J. (1996) Agricultural power in the European Community. Explaining the fate of the CAP and GATT negotiations. *Comparative Politics* 28 (2), 127–149.

Kjeldahl, R. (1994) Reforming the reform? – the CAP at a watershed. In: Kjeldahl, R. and Tracy, M. (eds), *Renationalisation of the Common Agricultural Policy?* Institute of Agricultural Economics, Copenhagen, pp. 5–22.

Koester, U. and Tangermann, S. (1977) Supplementing farm price support by direct payments. *European Review of Agricultural Economics* 4 (1), 7–31.

Kramer, R. and Batie, S. (1985) Cross compliance concepts in agricultural programs: the New Deal to the present. *Agricultural History* 59, 307–319.

Kuch, P. and Reichelderfer, K. (1992) The environmental implications of agricultural support programs: a United States perspective. In: Becker, T., Gray, R. and Schmitz, A. (eds), *Improving Agricultural Trade Performance under the GATT*. Wissenschaftverlag Vauk Kiel KG, pp. 215–231.

Larson, W., Pierce, F. and Dowdy, R. (1983) The threat of soil erosion to long term crop production. *Science* 219 (No. 4548), 458–464.

Leonard, P. (1982) Management agreements: a tool for conservation. *Journal of Agricultural Economics* 23 (3), 351–360.

Leuck, D. and Haley, S. (1996) Trade implications of the EU Nitrate Directive: An emerging research priority. In: Bredahl, M., Ballenger, N., Dunmore, J. and Roe, T. (eds), *Agriculture, Trade and the Environment: Discovering and Measuring the Critical Linkages*. Westview Press, Boulder, Colorado, pp. 231–242.

Levins, R. and Cochrane, W. (1996) The treadmill revisited. *Land Economics* 72 (4), 550–553.

Lovejoy, S. and Napier, T. (eds) (1986) *Conserving Soil: Insights from Socioeconomic Research*. Soil Conservation Society of America, Ankeny.

Lowe, P., Cox, G., O'Riordan, T. and Winter, M. (1986) *Countryside Conflicts: The Politics of Farming, Forestry and Conservation*. Temple Smith, London.

Lowe, P., Ward, N. and Munton, R. (1990) Social analysis of land use change: the role of the farmer. In: Whitby, M. (ed.), *Land Use Change: the Causes and Consequences*. ITE Symposium, HMSO, London, pp. 42–51.

Lowe, P., Ward, N., Ward, J. and Murdoch J. (1995) *Countryside Prospects 1995–2000: Some Future Trends*. Centre for Rural Economy Research, Newcastle.

Land Use Consultants (1995) *Countryside Stewardship Monitoring and Evaluation. Third Report*. LUC, London.

Lugar, R. (1994) The CRP in a time of change. In: *When Conservation Reserve Program Contracts Expire: the Policy Options*. Soil and Water Conservation Society, Ankeny, pp. 12–14.

Luick, R. (1996) High natural value farming in the Black Forest. In: Mitchell, K. (ed.), *The Common Agricultural Policy and Environmental Practices: Proceedings of a Seminar*. Institute for European Environmental Policy, London, pp. 20–33.

MacEwen, A. and MacEwen, M. (1982) *National Parks: Conservation or Cosmetics?* Allen & Unwin, London.

MacEwen, M. and Sinclair, G. (1983) *New Life for the Hills.* Council for National Parks, London.

MAFF (1992a) *The Pennine Dales: ESA Report of Monitoring 1991,* Report 1. HMSO, London.

MAFF (1992b) *The Broads: Environmentally Sensitive Area Report of Monitoring 1991,* Report 2. HMSO, London.

MAFF (1992c) *The Somerset Levels and Moors: Environmentally Sensitive Area Report of Monitoring 1991,* Report 3. HMSO, London.

MAFF (1992d) *The South Downs: Environmentally Sensitive Area Report of Monitoring 1991,* Report 4. HMSO, London.

MAFF (1992e) *West Penwith: Environmentally Sensitive Area Report of Monitoring 1991,* Report 5. HMSO, London.

MAFF (1992f) *Environment Now at the Centre of the CAP.* Press release 216/92, MAFF, London.

MAFF (1993a) *Suffolk River Valleys: Environmentally Sensitive Area Report of Monitoring 1992,* Report 6. HMSO, London.

MAFF (1993b) *The Test Valley: Environmentally Sensitive Area Report of Monitoring 1992,* Report 7. HMSO, London.

MAFF (1993c) *Breckland: Environmentally Sensitive Area Report of Monitoring 1992,* Report 8. HMSO, London.

MAFF (1993d) *The Shropshire Borders: Environmentally Sensitive Area Report of Monitoring 1992,* Report 9. HMSO, London.

MAFF (1993e) *The North Peak: Environmentally Sensitive Area Report of Monitoring 1992,* Report 10. HMSO, London.

MAFF (1995a) *European Agriculture: the Case for Radical Reform.* MAFF, London.

MAFF/DOE (1995b) *Environmental Land Management Schemes in England.* Consultation document. MAFF/DOE, London.

MAFF (1996a) *The Broads ESA: Report of Environmental Monitoring 1987–95.* MAFF, London.

MAFF (1996b) *The Pennine Dales ESA: Report of Environmental Monitoring 1987–95.* MAFF, London.

MAFF (1996c) *The Somerset Levels and Moors ESA: Report of Environmental Monitoring 1987–95.* MAFF, London.

MAFF (1996d) *South Downs ESA: Report of Environmental Monitoring 1987–95.* MAFF, London.

MAFF (1996e) *West Penwith ESA: Report of Environmental Monitoring 1987–95.* MAFF, London.

MAFF (1997) Evidence to House of Commons Select Committee on Agriculture, *Environmentally Sensitive Areas and Other Schemes under the Agri-Environmental Regulation,* Vol. II, Minutes of Evidence, Second Report, Session 1996–97. HMSO, London, pp. 1–49.

Majone, G. (1989) *Evidence, Argument and Persuasion in the Policy Process.* Yale University Press, New Haven, Connecticut.

Majone, G. (1992) *Ideas, Interests and Policy Change.* European University Institute Working Paper 92/21, European University Institute, Florence.

Malone, L. (1986) A historical essay on the conservation provisions of the 1985 Farm Bill: sodbusting, swampbusting, and the conservation reserve. *Kansas Law Review* 34, 577–597.

Margheim, G. (1994) Soil erosion and sediment control. In: *When Conservation Reserve Program Contracts Expire: The Policy Options.* Soil and Water Conservation Society, Ankeny, pp. 15–18.

McCalla, A. (1993) Agricultural trade liberalisation: the ever elusive grail. *American Journal of Agricultural Economics* 75 (5), 1102–1112.

McCracken, D. and Bignal, E. (eds) (1995) *Farming on the Edge: The Nature of Traditional Farmland in Europe.* Proceedings of the Fourth European Forum on Nature Conservation and Pastoralism, Joint Nature Conservation Committee, Peterborough.

McCrone, G. (1961) *The Economics of Subsidising Agriculture.* Allen & Unwin, London.

McHenry, H. (1996) Understanding farmers' perceptions of changing agriculture: some implications for agri-environmental schemes. In: Curry, N. and Owen, S. (eds), *Changing Rural Policy in Britain: Planning, Administration, Agriculture and the Environment.* The Countryside and Community Press, Cheltenham, pp. 225–243.

McMichael, P. (1993) World food system restructuring under a GATT regime. *Political Geography* 12 (3), 198–214.

Meeus, J., Wijermans, M. and Vroom, M. (1990) Agricultural landscapes in Europe and their transformation. *Landscape and Urban Planning* 18 (3/4), 289–352.

Merricks, P. (1990) *A Review of Environmentally Sensitive Areas,* Vol. I, Main Report. Nature Conservancy Council, Peterborough.

Metais, M. (1993) The role of ESAs in nature conservation in France. In: Dixon, J., Stones, J. and Hepburn, I. (eds), *A Future for Europe's Farmed Countryside, Studies in European Agriculture and Environmental Policy.* Royal Society for the Protection of Birds, Sandy, pp. 95–104

Ministry of Town and Country Planning (1947) *Conservation of Nature in England and Wales.* HMSO, London.

Miranowski, J. (1988) Monitoring the economic effects of the conservation reserve. *Journal of Soil and Water Conservation* 43 (1), 59–60.

Miranowski, J. and Reichelderfer, K. (1985) Resource conservation programs in the farm policy arena. In: *USDA Agricultural–Food Policy Review: Commodity Program Perspectives.* Economic Research Service, USDA, Washington, DC.

Miranowski, J., Hrubovcak, J. and Sutton, J. (1991) The effects of commodity programs on resource use. In: Just, R.E. and Bockstael, N.E. (eds), *Commodity and Resource Policies in Agricultural Systems.* Springer-Verlag, New York, pp. 202–221.

Moore, N. (1987) *The Bird of Time. The Science and Politics of Nature Conservation.* Cambridge University Press, Cambridge.

Morris, C. and Potter, C. (1995) Recruiting the new conservationists: farmers' adoption of agri-environmental schemes in the U.K. *Journal of Rural Studies* 11 (1), 51–63.

Moyer, W. and Josling, T. (1990) *Agricultural Policy Reform: Politics and Process in the EC and the USA.* Harvester Wheatsheaf, New York.

Munton, R. (1983) Agriculture and conservation: what room for compromise? In: Warren, A. and Goldsmith, F. (eds), *Conservation in Perspective.* John Wiley & Sons, Chichester, pp. 353–373.

Myers, P. (1988) Conservation at the crossroads. *Journal of Soil and Water Conservation* 43 (1), 10–13.

NCC (1977) *Nature Conservation and Agriculture.* Nature Conservancy Council, London.

NCC (1984) *A Nature Conservation Strategy for Great Britain.* Nature Conservancy Council, London.

Nelson, R. and Soete, L. (1988) Policy conclusions. In: Dosi, G., Freman, G., Silerberg, G. and Soete, L. (eds), *Technical Change and Economic Theory.* Pinter, London.

Neville-Rolfe, E. (1984) *The Politics of Agriculture in the European Community.* European Centre for Political Studies, Policy Studies Institute, London.

Nevin, E. (1990) *The Economics of Europe.* Macmillan, London.

Newby, H. (1993) The social shaping of agriculture: where do we go from here? *Journal of the Royal Agricultural Society of England* 154, 9–18.

NFU (1995) *Taking Real Choices Forward.* National Farmers' Union, London.

NFU (1996) Evidence to House of Commons on Agriculture, *Environmentally Sensitive Areas and Other Agri-Environmental Schemes,* Vol. II, *Minutes of Evidence.* Second Report, Session 1996–97. HMSO, London, pp. 79–81.

Nowicki, P. (1988) The CAP and the environment in France. In: Gilg, A. (ed.), *International Yearbook of Rural Planning.* Elsevier, London, pp. 325–337.

NSCGP (Netherlands Scientific Council for Government Policy) (1992) *Ground for Choices.* NSCGP, The Hague.

O'Riordan, T. (1984) The agriculture–conservation dispute: saturation coverage? In: Korbey, A. (ed.), *Investing in Rural Harmony: a Critique.* Centre for Agricultural Strategy, University of Reading, pp. 46–59.

O'Riordan, T. (1985) Halvergate: the policy and politics of change. In: Gilg, A. (ed.), *Countryside Planning Yearbook.* Geobooks, Norwich, pp. 101–106.

O'Riordan, T. and Bentham, G. (1993) The politics of nitrate in the UK. In: Burt, T., Heathwaite, A. and Trudgill, S. (eds), *Nitrate: Processes, Patterns and Management.* John Wiley & Sons, Chichester, pp. 403–416.

OECD (1995) *Agricultural Policies, Markets and Trade, Monitoring and Outlook Report.* OECD, Paris.

OECD (1997) *The Environmental Effects of Agricultural Land Diversion Schemes.* OECD, Paris.

Official Journal of the European Communities (1984*) Debates of the European Parliament.* Session 1983–84, 1–313, Brussels.

Offutt, S. (1996) Subsidising agriculture: the road ahead. *Choices,* Second Quarter 1996, 30–33.

Offutt, S. and Shoemaker, R. (1988) *Farm Programs Slow Technology-Induced Decline in Land's Importance.* Economic Research Service Technical Bulletin No. 1745. USDA, Washington, DC.

Ogg, C. and Zellner, J. (1984) A conservation reserve: conserving soil and dollars. *Journal of Soil and Water Conservation* 39 (2), 92–94.

Ogg, C., Webb, S-E. and Huang, W-Y. (1984) Cropland acreage reduction alternatives: an economic analysis of a soil conservation reserve and competitive bids. *Journal of Soil and Water Conservation* 39 (6), 379–383.

Ogg, C., Hostetler, J. and Lee, D. (1988) Expanding the conservation reserve to achieve multiple environmental goals. *Journal of Soil and Water Conservation* 43 (1), 78–81.

Olson, M. (1965) *The Logic of Collective Action.* Harvard University Press, Cambridge, Massachusetts.

Orden, D., Paarlberg, R. and Roe, T. (1996) A Farm Bill for booming markets. *Choices,* Second Quarter 1996, 13–16.

Osborn, C. (1993) The Conservation Reserve Program: status, future and policy options. *Journal of Soil and Water Conservation* 48 (4), 271–277.

Osborn, C. (1994) A national survey of CRP contract-holders. In: *When Conservation Reserve Program Contracts Expire: the Policy Options.* Soil and Water Conservation Society, Ankeny, pp. 60–63.

Osborn, C. and Miranowski, J. (1994) Lessons from U.S. policies for soil erosion control. In: *Agriculture and the Environment in the Transition to a Market Economy.* OECD, Paris, pp. 221–230.

Osborn, C., Schnepf, M. and Keim, R. (1994) *The Future Use of Conservation Reserve Program Acres: a National Survey of Farm Owners and Operators.* Soil and Water Conservation Society, Ankeny.

OTA (1995a) *Agriculture, Trade, and Environment: Achieving Complementary Policies.* Office of Technology Assessment, US Congress, Washington, DC.

OTA (1995b) *Targeting Environmental Priorities in Agriculture: Reforming Program Strategies.* Office of Technology Assessment, US Congress, Washington, DC.

Paarlberg, D. (1988) *Farm and Food Policy: Issues of the 1980s.* University of Nebraska Press, Lincoln, Nebraska.

Paarlberg, R. (1989) Is there anything 'American' about American agricultural policy? In: Kramer, C. (ed.), *The Political Economy of U.S. Agriculture: Challenges for the 1990s.* Resources for the Future, Washington, DC, pp. 37–55.

Pain, D. (1994) Case studies of farming and birds in Europe: Transhumance pastoralism in Spain. Unpublished research report, Royal Society for the Protection of Birds, Sandy.

Pain, D. and Pienkowski, M. (eds) (1997) *Farming and Birds in Europe.* Academic Press, London.

Pearce, D., Markandya, A. and Barbier, E. (1989) *Blueprint for a Green Economy.* Earthscan, London.

Petit, M. (1985) *Determinants of Agricultural Policies in the United States and the European Community.* Research Report 51, International Food Policy Research Institute, Washington, DC.

Phillips, P. (1990) *Wheat, Europe and the GATT: a Political Economy Analysis.* Pinter Publishers, London.

Phipps, T. and Reichelderfer. K. (1989) Farm support and environmental quality at odds? *Resources* 95, 20–33.

Phipps, T., Rossmiller, G. and Myers, W. (1990) Decoupling and related farm policy options. In: Allen, K. (ed.), *Agricultural Policies in a New Decade.* Resources for the Future, Washington, DC, pp. 101–123.

Porchester (1977) *A Study of Exmoor.* Report to the Secretary of State for the Environment and Minister of Agriculture, Fisheries and Food. HMSO, London.

Potter, C. (1983) *Investing in Rural Harmony: An Alternative Package of Agricultural Subsidies and Incentives for England and Wales.* World Wildlife Fund, Godalming.

Potter, C. (1986) Processes of countryside change in lowland England. *Journal of Rural Studies* 2 (3), 187–195.

Potter, C. (1988) Environmentally Sensitive Areas: an experiment in countryside management. *Land Use Policy* 5 (3), 301–313.

Potter, C. (1990) Conservation under a European farm survival policy. *Journal of Rural Studies* 6 (1), 1–7.

Potter, C. (1995) Environmental problems in agriculture and their underlying causes. Unpublished paper presented to OECD experts workshop on economic instruments for achieving environmental goals in the agricultural sector, 10 June, Paris.

Potter, C. (1996) *Decoupling by Degrees: The Nature Conservation Effects of Agricultural Liberalisation.* English Nature Research Report 196, English Nature, Peterborough.

Potter, C. (1997) Europe's changing farmed landscapes. In: Pain, D.J. and Pienkowski, M.W. (eds), *Farming and Birds in Europe.* Academic Press, London, pp. 25–41.

Potter, C. and Lobley, M. (1993) Helping small farmers and keeping Europe beautiful. *Land Use Policy,* October 1993, 267–279.

Potter, C. and Lobley, M. (1996) The farm family life cycle, succession paths and environmental change in Britain's countryside. *Journal of Agricultural Economics* 47 (2), 172–190.

Potter, C., Burnham, P., Edwards, A., Gasson, R. and Green, B. (1991) *The Diversion of Land: Conservation in a Period of Farming Contraction.* Routledge, London.

Potter, C., Barr, C. and Lobley, M. (1996) Environmental change in Britain's countryside: an analysis of recent patterns and socio-economic processes based on the Countryside Survey 1990. *Journal of Environmental Management* 48 (2), 169–186.

Primdahl, J. (1996) Denmark. In: Whitby, M. (ed.), *The European Environment and CAP Reform: Policies and Prospects for Conservation.* CAB International, Wallingford, pp. 45–69.

Rasmussen, W. (1982) History of soil conservation, institutions and incentives. In: Halcrow, H., Heady, E. and Cotner, M. (eds), *Soil Conservation Policies, Institutions, and Incentives.* Soil Conservation Society of America, Ankeny, pp. 3–18.

Rasmussen, W. and Baker, G. (1972) *The Department of Agriculture.* Praeger Press, New York.

Rausser, G. (1990) Political economic markets: PERTS and PESTS in food and agriculture. *American Journal of Agricultural Economics* 64 (5), 821–833.

Rausser, G. and Foster, W. (1989) The evolution and coordination of U.S. commodity and resources policies, a coherent policy for US agriculture. In: Horowich, G. and Lynch, G. (eds), *Food, Policy and Politics: A Perspective on Agriculture and Development.* Westview Press, Boulder, Colorado, pp. 191–237.

Rausser, G. and Irwin, D. (1988) The political economy of agricultural policy reform. *European Review of Agricultural Economics* 15 (4), 349–366.

Rayner, A., Ingernsent, K. and Hine, R. (1993) Agriculture in the Uruguay Round: an assessment. *Economic Journal* 103, 1513–1527.

Reeve, R. (1993) Making environmental policy: Britain, the Netherlands and the new EC nitrates directive. In: Bolsius, E., Clark, G. and Groenendijk, J. (eds), *The Retreat: Rural Land-Use and European Agriculture.* Royal Dutch Geographical Society, Utrecht, pp. 61–81.

Reichelderfer, K. (1985) *Do USDA Farm Program Participants Contribute to Soil Erosion?* Economic Research Service, USDA, Washington, DC.

Reichelderfer, K. (1990) Environmental protection and agricultural support. Are trade-offs necessary? In: Allen, K. (ed.), *Agricultural Policies in a New Decade.* Resources for the Future, Washington, DC, pp. 201–230.

Reichelderfer, K. (1992) Land stewards or polluters?: the treatment of farmers in the evolution of environmental and agricultural policy. In: Swanson, L. and Clearfield, F.

(eds), *Farming and the Environment: Iron Fist or Open Hand?* Soil and Water Conservation Society, Ankeny, pp. 15–28.

Reichelderfer, K. and Boggess, W. (1988) Government decision making and program performance: the case of the Conservation Reserve Program. *American Journal of Agricultural Economics* 70 (1), 1–11.

Reichelderfer, K. and Hinkle, M. (1989) The evolution of pesticide policy: environmental interests and agriculture. In: Kramer, C. (ed.), *The Political Economy of U.S. Agriculture: Challenges for the 1990s.* Resources for the Future, Washington, DC, pp. 147–173.

Reimenschneider, C. and Young, R. (1990) Agriculture and the failure of the budget process. In: Kramer, C. (ed.), *The Political Economy of U.S. Agriculture: Challenges for the 1990s.* Resources for the Future, Washington, DC, pp. 87–102.

Ribaudo, M. (1989) Targeting the Conservation Reserve Program to maximise water quality benefits. *Land Economics* 65 (4), 320–331.

Ribaudo, M., Piper, P., Schaible, G., Langner, L. and Colacicco, D. (1989) CRP: what economic benefits? *Journal of Soil and Water Conservation* 44 (5), 421–424.

Ribaudo, M., Osborn, C. and Konyar, K. (1994) Land retirement as a tool for reducing agricultural nonpoint source pollution. *Land Economics* 70 (1), 77–87.

Ringqvist, E. (1993) *Environmental Protection at the State Level: Politics and Progress in Controlling Pollution.* Sharp, London.

Rominger, R. (1994) The CRP's niche in the administration's environmental agenda. In: *When Conservation Reserve Program Contracts Expire: the Policy Options.* Soil and Water Conservation Society, Ankeny, pp. 53–56.

Ronningen,V. and Dixit, P. (1991) *A Single Measure of Trade Distortion.* IARTC Working Paper, USDA, Washington, DC.

Rosenthal, G. (1975) *The Men Behind Decisions: Cases in European Policy-Making.* D.C. Heath and Co., Lexington.

Rude, S. and Frederiksen, B. (1994) *National and EC Nitrate Policies – Agricultural Aspects for Seven EC Countries.* Statens Jordbrugsøkonomiske Institut, Copenhagen.

Rundqvist, B. (1996) Sweden. In: Whitby, M. (ed.), *The European Environment and CAP Reform: Policies and Prospects for Conservation.* CAB International, Wallingford, pp. 173–185.

Runge, F. (1988) The assault on agricultural protectionism. *Foreign Affairs*, Autumn 1988, 133–150.

Runge, F. (1994) *Designing Green Support: Incentive Competition and the Conservation Programs.* Center for International Food and Agricultural Policy, University of Minnesota, St Paul, Minnesota.

Russell, N. and Froud, J. (1991) *Socio-Economic Monitoring of the North Peak and Suffolk River Valleys Environmentally Sensitive Areas Scheme.* Ministry of Agriculture, Fisheries and Food, London.

Sabatier, P. (1987) Knowledge, policy oriented learning and policy change. An advocacy coalition framework. *Knowledge: Creation, Diffusion, Utilization* 8 (4), 64–92.

SAFE (1996) *Agri-Environmental Schemes in Europe – a Project of the European Network of Alliances for Sustainable Agriculture, Survey Report and Symposium Proceedings.* SAFE Alliance, London.

Scheele, M. (1996) The agri-environmental measure in the context of CAP reform. In: Whitby, M. (ed.), *The European Environment and CAP Reform: Policies and Prospects for Conservation.* CAB International, Wallingford, pp. 3–7.

Scott, Lord Justice (1942) *Committee on Land Utilization in Rural Areas.* Cmd 6378, HMSO, London.

Selman, P. (1993) Landscape ecology and countryside planning: vision, theory and practice. *Journal of Rural Studies* 9 (1), 1–21.

Seymour, S., Cox, G. and Lowe, P. (1992) Nitrates in water: the politics of the 'polluter pays principle'. *Sociologia Ruralis* 22 (1), 82–103.

Sheail, J. (1985) *Pesticides and Nature Conservation: the British Experience, 1950–1975.* Clarendon Press, Oxford.

Sheail, J. (1995a) Nature protection, ecologists and the farming context: a UK historical context. *Journal of Rural Studies* 11 (1), 79–88.

Sheail, J. (1995b) War and the development of nature conservation in Britain. *Journal of Environmental Management* 44 (7), 267–283.

Sheperd, L. (1985) The farm debt crisis: temporary or chronic? Paper presented at the annual meeting of the Western Economics Association, Anaheim, California.

Shoard, M. (1980) *The Theft of the Countryside.* Temple Smith, London.

Slangen, L. (1992) Policies for nature and landscape conservation in Dutch agriculture: an evaluation of objectives, means, effects and programme costs. *European Review of Agricultural Economics* 19 (3), 331–350.

Smith, M. (1990) *The Politics of Agricultural Support in Britain.* Dartmouth Press, Aldershot.

Stanners, D. and Bordeau, P. (1995) *Europe's Environment: the Dobris Assessment.* Office for Official Publications of the European Communities, Luxembourg.

Suárez, F., Naveso, M. and De Juana, E. (1996) Farming in the drylands of Spain: birds of the pseudosteppes. In: Pain, D. and Pienkowski, M. (eds), *Farming and Birds in Europe.* Academic Press, London, pp. 297–330.

Swanson, L. (1989) Commentary on Chapter 4. In: Kramer, C.S. (ed.), *The Political Economy of U.S. Agriculture: Challenges for the 1990s.* Resources for the Future, Washington, DC, pp. 80–83.

Swanson, L. (1993) Agro-environmentalism: the political economy of soil erosion in the USA. In: Harper, S. (ed.), *The Greening of Rural Policy: International Perspectives.* Belhaven Press, London, pp. 99–118.

SWCS (1994) *When Conservation Reserve Program Contracts Expire: the Policy Options.* Soil and Water Conservation Society, Ankeny.

Swinbank, A. (1993) CAP Reform, 1992. *Journal of Common Market Studies* 31 (3), 359–372.

Swinbank, A. (1996) Capping the CAP? Implementation of the Uruguay Round agreement by the European Union. *Food Policy* 21 (4/5), 393–407.

Swinbank, A. and Tanner, C. (1996) *Farm Policy and Trade Conflict: the Uruguay Round and CAP Reform.* University of Michigan Press, Ann Arbor, Michigan.

Taft, S. and Runge, C. (1987) Supply control, conservation, and budget restraint: conflicting instruments in the 1985 farm bill. In: Halbach, D., Runge, C. and Larson, W. (eds), *Making Soil and Water Conservation Work: Scientific and Policy Perspectives.* Soil Conservation Society of America, Ankeny, pp. 3–13.

Tamminga, G. and Wijnands, J. (1991) Animal waste problems in the Netherlands. In: Hanley, N. (ed.), *Farming and the Countryside: an Analysis of External Costs and Benefits.* CAB International, Wallingford, pp. 117–136.

Tangermann, S. (1992) *Reforming the CAP? In for a Penny, In for a Pound.* IEA Inquiry 28, Institute of Economic Affairs, London.

Tangermann, S. (1996) Implementation of the Uruguay Round agreement on agriculture: issues and prospects. *Journal of Agricultural Economics* 47 (3), 315–337.

Tangermann, S. and Josling, T. (1995) *Towards a CAP for the Next Century*. European Policy Forum, London.

Thigpen, J. (1994) Soil conservation influences: A historical approach. In: Swanson, L. and Clearfield, F. (eds), *Agricultural Policy and the Environment: Iron Fist or Open Hand?* Soil and Water Conservation Society, Ankeny, pp. 65–80.

Thomas, K. (1983) *Man and the Natural World: Changing Attitudes in England, 1500–1800.* Allen Lane, London.

Thomson, K. (1996) The CAP and the WTO after the Uruguay Round Agriculture Agreement. Unpublished paper prepared to CREDIT conference, CAP Reform: What Next? CREDIT, University of Nottingham.

Tiner, R. (1984) *Wetlands of the United States: Current Status and Recent Trends.* Fish and Wildlife Service, Department of the Interior, Washington, DC.

Tobey, J. and Reinert, K. (1991) The effects of domestic agricultural policy reform on environmental policy. *The Journal of Agricultural Economics Research* 43 (2), 20–28.

Tracy, M. (1985) The decision-making practice of the European Community. In: Pelkmans, J. (ed.), *Can the CAP be Reformed?* European Institute of Public Administration, Maastricht, pp. 79–93.

Tracy, M. (1989) *Government and Agriculture in Western Europe, 1880–1988.* Harvester Wheatsheaf, Hemel Hempstead.

Turner, K. (1980) *The Impact of Modern Farming Systems on the Social and Physical Environment.* Farmer's Club, London.

Tutwiler, M. (1996) *Agra-Europe News Report, October 1996.* Agra-Europe, Tunbridge Wells, Kent.

Tweeten, L. (1979) *Foundations of Farm Policy.* University of Nebraska Press, Lincoln, Nebraska.

Ufkes, F. (1993) Trade liberalization, agro-food politics and the globalization of agriculture. *Political Geography* 12 (3), 198–214.

USDA (1981) *National Summary, Evaluation of the Agricultural Conservation Program: Phase 1.* USDA, Washington, DC.

USDA (1985) *Agricultural-Food Policy Review: Commodity Program Perspective.* Economic Research Service, USDA, Washington, DC.

USDA (1989) *Agricultural Statistics.* USDA, Washington, DC.

USDA (1990) *Status Review of Conservation Plans.* Audit Report, USDA, Washington, DC.

USDA (1993) *Conservation Reserve Program Enrolment Statistics, Sign-Ups 1–12.* USDA, Washington, DC.

US Department of the Interior (1988) *The Impact of Federal Programs on Wetlands.* Report to Congress, Washington, DC.

US Environmental Protection Agency (1990a) *National Pesticide Survey.* EPA, Washington, DC.

US Environmental Protection Agency (1990b) *Pesticides in Groundwater Database: a Compilation of Monitoring Studies.* EPA, Washington, DC.

Vail, D., Hasund, K. and Drake, L. (1994) *The Greening of Agricultural Policy in Industrial Countries: Swedish Reforms in Comparative Perspective.* Cornell University Press, Ithaca, New York.

van der Bijl, G. and Ooserveld, E. (1996) Netherlands. In: Whitby, M. (ed.), *The European Environment and CAP Reform: Policies and Prospects for Conservation.* CAB International, Wallingford, pp. 155–172.

von Meyer, H. (1988) *The Common Agricultural Policy and the Environment – the Effects of Price Policy and Options for its Reform.* CAP Discussion Paper No. 1, WWF International, Gland.

von Meyer, H. (1993) Agriculture and the environment in Europe. In: Group of Sesimbra (eds), *The European Common Garden: Towards the Building of a Common Environmental Policy.* Group of Sesimbra, Brussels, pp. 70–85.

Ward, N., Marsden, T. and Munton, R. (1990) Farm landscape change: trends in upland and lowland Britain. *Land Use Policy* 7 (4), 291–302.

Waters, G. (1994) Government policies for the countryside. *Land Use Policy* 11 (2), 88–93.

Weale, A. (1992) *The New Politics of Pollution.* Manchester University Press, Manchester.

Webster, S. and Felton, M. (1993) Targeting for nature conservation in agricultural policy. *Land Use Policy* 10 (19), 67–82.

Westmacott, R. (1983) The conservation of farmed landscapes: attitudes and problems in the United States and Britain. *Landscape Design* 144, 11–14.

Whitbread, A. and Jenman, W. (1995) A natural way of conserving biodiversity in Britain. *British Wildlife* 7 (2), 84–93.

Whitby, M. (1994) What future for ESAs? In: Whitby, M. (ed.), *Incentives for Countryside Management.* CAB International, Wallingford, pp. 253–271.

Whitby, M. (1996) The United Kingdom. In: Whitby, M. (ed.), *The European Environment and CAP Reform: Policies and Prospects for Conservation.* CAB International, Wallingford, pp. 186–205.

Whitby, M. and Harvey, D. (1988) Issues and policies. In: Ollerenshaw, J. and Whitby, M. (eds), *Land Use and the European Environment.* Belhaven, London, pp. 143–177.

Whitby, M. and Lowe, P. (1994) The political and economic roots of environmental policy in agriculture. In: Whitby, M. (ed.), *Incentives for Countryside Management.* CAB International, Wallingford, pp. 1–24.

Whitby, M. and Saunders, C. (1994) *Estimating the Supply of Conservation Goods.* Centre for Rural Economy Working Paper, University of Newcastle on Tyne.

Whitby, M., Coggins, G. and Saunders, C. (1990) Alternative payment systems for management agreements. Unpublished report to the Nature Conservancy Council, Peterborough.

Wildlife Trusts (1996) *Crisis in the Hills.* Wildlife Trusts, Lincoln.

Wilkinson, W. (1986) Press release, Nature Conservancy Council, London.

Wilkinson, A., Bernstein, N., Delorme, H., Henericks, G., Berkhout, P., Meester, G. and Nedergaard, P. (1994) Renationalisation: an evolving debate. In: Kjeldahl, R. and Tracy, M. (eds), *Renationalisation of the Common Agricultural Policy?* Institute of Agricultural Economics, Copenhagen, pp. 21–32.

Wilson, G.A. (1994) German agri-environmental schemes. A preliminary review. *Journal of Rural Studies* 10 (1), 27–45.

Wilson, G.A. (1997) Selective targeting in Environmentally Sensitive Areas: Implications for farmers and the environment. *Journal of Environmental Planning and Management* 40 (2), 199–215.

Wilson, G.K. (1977) *Special Interests and Policymaking. Agricultural Policies and Politics in Britain and the United States.* John Wiley & Sons, London.

Winter, M. (1997) CAP reform and the environment. In: Rayner, A., Ingersent, K. and Hine, R. (eds), *The Reform of the CAP*. Macmillan, London, in press.

Winters, L. (1987) The political economy of the agricultural policy of industrial countries. *European Review of Agricultural Economics* 14 (3), 285–304.

Winters, L. (1990) The road to Uruguay. *The Economic Journal* 100 (4), 1288–1303.

Woods, A., Taylor, J., Harley, D., Houseden, S. and Lance, A. (1988) *The Reform of the Common Agricultural Policy: New Opportunities for Wildlife and the Environment*. Royal Society for the Protection of Birds, Sandy.

Yearley, S., Mittom, K. and Baker, S. (1994) *Protecting the Periphery: Environmental Policy in the Peripheral Regions of the EU*. Frank Goss, Ilford.

Young, C. and Osborn, C. (1990a) *The Conservation Reserve Program: an Economic Assessment*. Economic Research Service 626, USDA, Washington, DC.

Young, C. and Osborn, C. (1990b) Costs and benefits of the Conservation Reserve Program. *Journal of Soil and Water Conservation* 45 (3), 370–373.

Young, D., Walker, D. and Kenjo, P. (1991) Cost effectiveness and equity aspects of soil conservation programs in a highly erodible region. *American Journal of Agricultural Economics* 73 (4), 1053–1062.

Zinn, J. (1988) *The Conservation Reserve: a Status Report*. Report No. 88-716 ENR, Congressional Research Service, Washington, DC.

Zinn, J. (1991) Conservation in the 1990 Farm Bill: the revolution continues. *Journal of Soil and Water Conservation* 46 (1), 45–48.

Zinn, J. (1993) How are soil erosion control programs working? *Journal of Soil and Water Conservation* 48 (4), 254–258.

Zinn, J. (1994) Resource conservation entering the 21st century: How will it change? In: Swanson, L. and Clearfield, F. (eds), *Agricultural Policy and the Environment: Iron Fist or Open Hand?* Soil and Water Conservation Society, Ankeny, pp. 29–46.

Zulauf, C., Tweeten, L. and Lines, A. (1996) The Federal Agricultural Improvement and Reform Act (FAIR): selected implications and unanswered questions. *Choices*, Second Quarter, 40–41.

Index

abandonment of land 17, 30, 32, 34, 43, 108, 115, 117, 123, 126, 142, 143, 145, 149

Accompanying Measures 101, 116, 117–118, 122, 137, 154, 157

Acreage Reduction Program (ARP) 14, 20, 70–71, 141

'additionality' 89, 92, 126

afforestation 30, 68, 86

Aggregate Measure of Support (AMS) 132, 133, 136–137

Agricultural Adjustment Act (US) (1933) 9–10, 37

Agricultural Adjustment Administration (AAA) 37

Agricultural Conservation Program (ACP) 37–38, 40, 42

agricultural decline 6–7, 108, 149

agricultural interest groups 11, 146, 150, 155

agricultural protectionism 10, 14, 16, 53, 129, 130, 131

Agricultural Stabilization and Conservation Service (ASCS) 38, 40, 41, 42, 50, 64, 67

agricultural trade liberalization 7, 57, 128–154, 160–162

Agriculture Act (UK) (1947) 14

Agriculture Act (US) (1973) 14

Agriculture Council (EU) see Council of Agriculture Ministers

Agri-Environmental Policy 108, 116–123, 126–129, 145, 147–148, 152, 154, 156, 157–158
  likely evolution of 157–162

Agri-Environmental Programme (AEP)
  EU budget 118, 121, 123, 124, 129, 157, 159
  in general 152, 154, 156–157, 159
  in the EU 3, 5, 6, 101, 105, 114, 115–127, 145, 146, 149, 151
  in the US 148
  see also Agri-Environmental schemes and Regulation 2078/92

agri-environmental schemes
  EU: after AEP establishment 106–108, 109–110, 118–122, 123–126, 128
  EU: before AEP establishment 51, 82–92, 93–97, 98, 99, 100, 101–104

agri-environmentalists 5, 6, 24, 35, 50, 51, 57, 59, 67, 72, 73, 75, 81, 92, 96, 102, 122, 146, 155, 162

Alps 109

American Farm Bureau (AFB) 40, 50

American Farmland Trust (AFT) 40, 55, 56, 64, 78, 79, 146

Andalucia 31
Appennines 32
Arable Area Payment Scheme (AAPS)
    100, 103, 136
Australia 139
Austria 118, 120, 121, 125

Baden-Württemberg 113, 143
Barlow, Tom 42
base acreage 22, 41, 55, 64, 70,
    see also base hectarage
'base bite' 64, 72, 74
base hectarage 74, *see also* base acreage
Bavaria 107
Belgium 15, 29
biodiversity 2, 25, 26, 29, 30, 31, 32,
    34, 35, 91, 97, 144, 152, 157
Biodiversity Convention 157
Blair House agreement 133, 137
Block, John 56
'blue baby syndrome' 111
'blue box' 133, 139
Brannan Plan (US) (1948) 12, 16
Brittany 29
Broads Grazing Marsh Scheme (BGMS)
    49–50, 58, 82–84, 85, 91

Cairns Group 132, 136
Castilla y Leon 30
Castro Verde Programme 122
Causse Mejan 143
Central and Eastern European Countries
    (CEEC) 137–138
Codes of Good Agricultural Practice 109,
    112, 113, 117, 148, 153
Committee of Professional Agricultural
    Associations (COPA) 17
commodity price supports *see* price
    support (US)
commodity support programmes *see* price
    support (US)
Common Agricultural and Rural Policy of
    Europe (CARPE) 151
Common Agricultural Policy (CAP)
    antagonism with agri-environmental
        measures 96–97, 103, 123,
        125, 154
    budget 127, 137

budget crisis 53–54, 59, 128
    effect on farm prices 16
    inception of 15–26
    likely future reform of 158,
        159–162
    reform of 2, 6–7, 8, 16–17, 45,
        57–60, 84, 96–97, 100–101,
        104, 105, 115, 117, 122, 129,
        135, 138, 140, 145, 152,
        156–157
    role in agricultural intensification
        28, 125
    role in agricultural specialization
        27, 96
    role in environmental change 2,
        27–34, 36, 43–45, 46, 48, 125,
        see also 'policy thesis'
    role in restructuring 26, 28, 105,
        130
compensation payments
    EU 100, 125, 136, 137, 139, 140,
        151, 161
    UK (SSSI) 47, 83
competitive bidding
    UK (CSS) 94, 95
    US (CRP) 62, 63, 71
'complementary technologies' 80, 148,
    149
conservation compliance
    EU adoption of 98–99, 101, 104,
        150, 151, 159
    US *see* Food Security Act (US) (1985)
    see also cross-compliance
conservation easements 74
Conservation Foundation 24, 41, 79
Conservation Reserve 62, 64–66, 68, 69,
    70, 71, 72, 155
Conservation Reserve Program (CRP)
    see Food Security Act (US) (1985)
Conservation Title *see* Food Security Act
    (US) (1985)
Cork Conference 151
Council of Agriculture Ministers (EU) 54
Council of Environmental Advisors to the
    Federal Government (CEAFG) 44
Council of Europe 29
Council for National Parks (UK) 48
Council for the Protection of Rural
    England (CPRE) 43, 47, 57, 59,
    92

Country Landowner's Association (CLA)
58, 59
Countryside Access Scheme (CAS)
101–102
countryside character 91, 103, 143,
151, 159
Countryside Commission (CC) 6, 28, 49,
58, 82, 85, 98, 103
Countryside Council for Wales (CCW) 95
countryside management 45, 49, 83, 84,
87, 88, 91, 95, 97, 110, 115,
117, 126, 127, 145, 152, 156,
157, 159
Countryside Premium Scheme (CPS) 98,
100, 102
'countryside products' 94, 95
Countryside Stewardship Scheme (CSS)
93–96, 97, 103, 104, 121
inception of 93–95
Countryside Survey 1990 36, 97
cross-compliance 13, 42–43, 50
see also conservation compliance,
EU adoption of

decoupling 7, 12, 48, 57–58, 77, 87, 94,
99, 101, 126, 132, 133–153,
160, 161
moderate decoupling 151, 153
radical decoupling 150
weak decoupling 150
deficiency payments (US) 11, 14, 22–23,
42, 52, 64, 70, 72, 133, 134,
135
dehesa 31–32
Denmark 10, 28, 33, 44, 106, 107,
112, 113, 118, 120, 156, 159
Department of the Environment (DOE)
35, 59, 95
desertification *see* abandonment of land
Deutsche Bauenverband (DBV) 14
developing countries 52, 138, 139
DG VI 105, 106, 109–110
direct payments
EU 57, 58, 96, 137
US 12, 78
see also compensation payments
Directives (EU)
Drinking Water 111–112
Habitats and Species 115, 157

Less Favoured Areas (LFAs) 17, 30,
50, 58, 107, 109
expenditure on 17
greening of 97, 100
origins of 17
Nitrates 102, 112–113, 114, 140

early retirement schemes 17
EC co-financing 105, 116, 118,
122, 123
Economic Research Service (ERS) 63–64,
146
Engel's Law 11
English Heritage (EH) 85
English Nature (EN) *see* Nature
Conservancy Council
environmental goods 99, 126, 139, 145,
150
Environmental Land Management
Scheme (ELMS) 101
environmental payments 127, 145
Environmental Protection Agency (EPA)
24, 156
Environmental Quality Incentives
Program (EQIP) 147
environmental value for money 3, 6, 67,
91, 93, 127, 150
Environmentally Sensitive Areas (ESA)
assessment of 91–92, 93, 103, 104
early design 84–86
farmer response to 86, 88–91
government expenditure on 101,
102
implementation in the EU 101, 103,
106–109, 120
European Agricultural Guarantee and
Guidance Fund (EAGGF) 54, 137
European Commission (EC) 6, 7, 17, 53,
57, 58, 106, 108, 109, 110, 111,
112, 114, 115, 116, 118,
119–120, 123, 124, 126, 127,
149, 157, 159
European Rural Policy 149, 151, 153, 161
European Union (EU) 5, 6, 7, 15, 17, 18,
33, 51, 53–54, 82, 93, 100, 105,
108, 109, 110, 118, 125, 126,
129, 130, 132, 133, 135, 136,
137, 138, 140, 144, 146, 150,
151, 156, 157, 159, 161

eutrophication 24, 29, 33, 110
export levy 132
export subsidies 52, 55, 130, 131, 133, 136
extensification 7, 57, 109, 110, 114–115, 117, 118, 120, 121, 124, 127, 138, 139, 140, 141, 156, 157
extensification effect 139, 140–141, 148, 152
extensification scheme 97, 125
Extremadura 31

farm
    amalgamations 19, 28, 143
    crisis (US) 14, 51–52
    debt 51
    structural change 15, 16, 23, 26, 44, 48
farm lobby 2, 4, 5, 51
    EU 50, 57–58, 112, 127, 160
    US 56, 134, 135, 140, 155
farmer response (to agri-environmental schemes) 64, 86, 89–91, 95, 97, 103, 160
Federal Agricultural Improvement and Reform (Act) (FAIR) (US) (1996) 134–135, 137, 141, 147, 152, 158, 161
federal rollback 146, 148, 152
fertilizer use
    EU 27, 29, 91–92, 110, 111, 112, 113, 114, 115, 117, 124, 139, 140
    US 19–20
    see also pollution, pesticide and fertilizer
Financial Management Initiative (FMI) 91
Finland 118, 121, 122, 125
Food and Agriculture Organization (FAO) 10
FOEGA *see* EAGGF
Food Security Act (FSA) (1985) (US) 61–62, 73, 74, 134
    Conservation Title 60–82
        conservation compliance 42–43, 50–51, 56–57, 62, 63, 64, 68, 76, 77, 79, 80, 90, 147, 155, 157, 158 (EU adoption of) 98–99, 101, 104, 150, 151, 159
    Conservation Reserve Program (CRP) 62–80, 81
    Highly Erodible Land (HEL) sub-title 62–63, 66
    invention of 61
    lobbying for 55–57, 60
    policy significance of 66–67, 76, 80, 155, 159
France 10, 14, 15, 27, 28, 30, 33, 43, 97, 107–108, 109, 110–111, 113, 117, 118, 120, 125, 144, 151
free trade 11, 99, 144, 146, 147, 160
'freedom to farm' 77, 130, 135, 140, 147, 152
Freedom to Farm Bill 134–135
Friends of the Earth (FoE) 47, 92

Gemeinschaftsaufgabe Verbersserung der Agrastruktur und des Kustenschutzes (Germany) (GAK) 120, 125
General Accounting Office (GAO) 68–69
General Agreement on Tariffs and Trade (GATT) 77, 117, 129–133
General Agri-Environmental Protection Scheme (Finland) (GAEPS) 122, 125
Germany 10, 14, 15, 29, 33, 43–44, 106, 107, 111, 112, 113, 114, 118, 120, 125, 144, 151, 159
Great Plains (US) 20, 21, 26, 38–39, 63, 69, 71, 75
Greece 27, 33, 118, 120
Green Book (EC) 54, 57
'green box' 129, 135, 137, 139, 161
Green Europe 17, 108–109, 123, 127, 135, 149
green payments 58, 128, 139, 146, 147, 149, 162
    see also environmental payments
green recoupling 80, 99–100, 129, 139, 140, 144–151, 152, 153, 159, 162
Green Ticket (US) 41–42, 98
Gross Domestic Product (GDP) 15, 114, 116

Habitat Directive *see* Directive, Habitats and Species
habitat loss 11, 25, 26, 29, 30, 36, 46
Habitat Scheme 102
'halo effect' 92
headage payments 48, 136
hectarage control 10, 52, 78
hectarage payments 149, 161
hectarage reduction program 2, 13, 14, 20, 22, 55
    see also Acreage Reduction Program (ARP)
hedgerow loss 28–29, 97
High Natural Value (HNV) farming systems 30, 34, 126, 143, 145, 149, 152
    see also low intensity farming systems
Highly Erodible Lands (HEL) 40, 55, 62, 67, 74, 76, 81
Hill Livestock Compensatory Allowance (HLCA) 97, 100
House Agriculture Committee (US) 54, 56, 61, 62, 124
House of Commons Agriculture Committee (UK) 58, 59, 103

import levies 10, 14, 53
income support 5, 57, 64, 67, 71, 74, 81, 87, 106, 109, 110, 123, 124, 126, 127, 139, 145, 150, 154, 158
Institute for European Environmental Policy (IEEP) 57
Institute of Terrestrial Ecology (ITE) 36, 101
intensification
    EU 24, 27, 28–31, 33, 34, 44, 46, 48, 57, 85, 97, 110, 125–126
    US 19, 23, 41
intensive livestock 26, 32–33, 110, 113
Ireland 27, 106
island biogeography 46
Italy 15, 28, 33, 106

Japan 131, 139
joint economies 28

joint products 83
Jura 109

KULAP 107

land reclamation 20, 25, 30, 46, 86, 139
land use change 20, 21, 22, 26, 87, 92, 142, 151
land-saving technologies 19–20, 139
landscape change 1, 6, 24–25, 28–29, 34, 36
landscape protection 6, 49, 50, 105, 107, 126, 144
Less Developed Country (LDC) *see* developing countries
Less Favoured Area (LFA) 43, 48, 109
Less Favoured Area Directive *see* Directive (LFA)
livestock extensification 117, 118, 125
livestock grazing pressure 32
livestock intensification 30, 31
livestock payments 48, 49–50, 97, 100
livestock premiums *see* livestock payments
livestock unit (LU) 120, 125
lobbyists 2, 6, 35, 36, 44, 54, 59, 60, 73, 82, 92, 129, 148
low intensity farming systems 83, 120, 125
    see also High Natural Value farming systems
Lugar, Senator 134
Luxembourg 15

MacSharry reforms
    background to 57, 97, 100
    environmental impact of 100, 101, 122, 123–124, 125, 127, *see also* AEP
    policy significance of 101, 123, 127, 128, 136–137, 157
    see also Regulation 2078/92 *and* CAP (reform of)
MacSharry, Ray 136
'magic bullet' 61, 155
management agreements 46, 48, 49, 50, 85, 107

Mansholt Plan 16
Mansholt, Sicco 16, 17, 34, 53
marginal (economically) farmers 17, 19, 53, 87, 123, 142, 143
marginal agricultural production 109
marginal areas 118
marginal producers *see* marginal farmers
Marshall Plan 14
Massif Central 109, 120
maximum acceptable rental rate (MARR) 71
'middle countryside' 144
milk quotas 54, 136
Ministry of Agriculture, Fisheries and Food (MAFF) (England) 6, 47, 50, 58–59, 82, 84, 86, 87–88, 89, 92, 93, 96, 98, 100, 103, 104, 112, 114, 120
Montados 31
Moorland Scheme (UK) 102

National Association of Conservation Districts (NACD) 42, 63
National Farmers' Union (NFU)
England and Wales 50, 84, 112
US 56
National Nature Reserve (NNR) 45
National Resources Inventory (NRI) 20–21, 23–24, 35, 38, 55, 146
National Soil Conservation Plan (US) 146
National Wetlands Forum (US) 26, 72
Natura 2000 124
see also Directive, Habitats and Species
Natural Resources Defence Council (US) 42
Nature Conservancy Council (NCC) 28, 29, 36, 45, 46, 47, 85
The Netherlands 10, 15, 26, 29, 32, 44, 58, 106, 107, 109, 110, 111, 112, 113, 114, 140–141, 156, 159
Netherlands Scientific Council for Government Policy (NSCGP) 144
New Deal 6, 10, 12, 20, 34, 36–37, 56, 57, 81, 130, 156
Nitrate Advisory Area (NAA) 112
nitrate pollution *see* pollution, nitrate
Nitrate Sensitive Area (NSA) 102–103, 112, 114

Nitrate Vulnerable Zone (NVZ) 102, 112, 113, 114
Non-Governmental Organization (NGO) 120
Norfolk Broads 49
see also Broads Grazing Marsh Scheme (BGMS)
Normandy 29
North American Free Trade Agreement (NAFTA) 141
Norway 10

Objective 1 areas (EU) 116
Office of Technology Assessment (OTA) 23, 24, 25, 80
olive groves 30, 31, 118
ÖPUL (national environment) scheme 122
'orange ticket' compliance 99
Organic Aid Scheme
Germany 120
UK 103
organic farming 117, 118, 122
Organization for Economic Co-operation and Development (OECD) 129, 131
Organization for European Economic Co-operation (OEEC) 15
over-grazing 31, 32, 100, 103, 139
over-production 5, 10, 12, 14, 155
control of 51, 60, 127, 132, 154
environmental impact 7

Payment in Kind (PIK) 14, 41, 131
Peace Clause 133
see also URAA
permanent crops 30, 31
pesticide use 6, 27, 72, 111, 140
see also pollution, pesticide and fertilizer
'policy thesis' 44, 59
polluter pays principle (PPP) 109, 113, 114, 125, 161
pollution
agricultural 6–7, 24, 33, 34, 43, 72, 110–115, 127, 147, 156, 158

ground water 4, 33, 139
manure 33, 111, 113
nitrate 24, 33, 43, 110, 111, 112,
118, 140
nitrogen 33, 122, 141
pesticide and fertilizer 22, 23, 24,
29, 110, 118, 141
water 35, 42, 69, 74, 113, 139,
147, 148
'pork barrel' politics 39
Portugal 30, 31, 118, 120
price support 7, 131, 132, 133
environmental impact of reduction in
21, 22, 23, 40
under CAP 5, 11, 14, 15, 16, 17,
29, 47, 53, 57, 97, 123, 138,157
under US Commodity Programs 9,
12, 13, 19, 20, 22, 23, 34, 36,
37, 38, 41, 50, 51, 52, 53, 56,
64, 76, 77, 79, 134, 135, 156
see also double dividend
prime à l'herbe (France) 120, 125
Producer Entitlement Guarantee (PEG)
151
Producer Subsidy Equivalent (PSE)
131
property rights 147
public goods 2, 7, 91, 101, 126, 139,
155, 156, 157
Pyrenees 109

quotas 57, 136

rare breeds 117, 120
regulation (environmental) 58, 72, 110,
111, 127, 148–149, 158, 159,
162
Regulations (EC)
797/85 (Article 19) 84, 105,
107–108, 109–110, 112,
115–117, 127
1760/87 97, 106, 114
2078/92 101, 116–125, 126, 157,
see also AEP
746/96 116, 159
rent-seeking 146, 150
Resource Conservation Appraisal (RCA)
21, 35, 39–40, 42–43, 56, 57

rural depopulation 17, 117
Rural Environment Protection Scheme
(REPS) (Ireland) 120

Scandinavia 3, 30
Schleswig-Holstein 107
Senate Agriculture Committee (US) 54,
56, 61, 62, 134
set-aside
environmental impact of
(US) 23, 155
(EU) 114–115
European adoption of 53, 97–98,
100, 102, 114–115, 117, 136,
137
US experience with 13, 23, 42, 65,
155
Shoard, Marion 1, 36
Single European Act (SEA) 110, 111
Site of Special Scientific Interest (SSSI)
45, 46, 58, 86
sodbuster measures 57, 62, 63, 64, 76,
77, 156
Soil and Water Conservation Society
(SWCS) 75, 77
Soil Bank Program 13, 38, 41, 68, 75
soil conservation programmes (US) 2,
35, 38, 39, 41, 42, 60, 63, 147,
152
Soil Conservation Service (SCS) 35, 38,
39, 42, 50, 66, 68, 76, 77, 146,
155
soil erosion
environmental impact of 20, 23–24,
66, 69, 72, 81, 158
measurement of 6, 21, *see also* USLE
trends in 1, 2, 20–24, 33, 34,
36–37, 39–41, 42, 55, 63, 65,
69, 74, 75, 76, 77, 81, 141, 147,
155, 156, 158
Spain 30–32, 33, 118, 122, 145
specialization (EU) 10–11, 27–28, 33,
44
stewardship (environmental) 1, 4, 8, 84,
88, 110, 116, 148, 155, 159,
161
Stresa Conference 17, 108
'structural learning' 88, 104
subsidiarity 124, 157

supply control
    EU 57, 106, 110, 132, 155
    US 63, 67, 70, 71, 72, 73, 81,
       155
swampbuster measures 63, 73, 74,
    76–77, 156
Sweden 10, 30, 120–121, 143, 159

tarification 14, 129, 130, 133, 136
technological change
    environmental impacts of 2, 10, 15,
      18, 20, 22, 33
    progress of 19
technological revolution *see* technological
    change
technological treadmill 19
'Theft of the Countryside' 1, 36
Tir Cymen (TC) (Wales) 94–95
transaction costs 19, 157
transhumance 32
Treaty of Rome (1958) 15, 53, 59, 156

under-grazing 32
United States Department of Agriculture
    (USDA) 2, 10, 13, 14, 19, 20,
    21, 23, 24, 35, 37, 39, 41, 43,
    55, 56, 61, 62–63, 64, 65, 67,
    68, 69, 71, 72, 74, 76, 79, 146,
    152, 155–156, 158, 161
Universal Soil Loss Equation (USLE)
    21, 23
Uruguay Round Agriculture Agreement
    (URAA) 7, 129, 131, 133, 136,
    137, 138, 139, 141, 147
US Congress 39, 40, 54, 57, 61, 63, 74,
    130, 134, 147

variable import levies 16, 130

water quality (US) 39, 68–69, 74
    (EU) 111
    deterioration in (US) 7
    improvement 146
    monetary valuation of
      66
    see also pollution, groundwater
      and water
wetlands
    conservation significance of in US
      26
    depletion of 25–26
    'no net loss' of 26, 72
Wetlands Reserve Program (WRP) 74,
    79, 147, 152
Wildlife and Countryside Act (WCA
    (1981) (UK)) 47, 49, 50,
    82
windfall gains 64
world food markets 10, 12, 14, 57, 78,
    99, 131
World Health Organization (WHO) 33
world prices 14, 53, 57, 131, 137, 138,
    144, 155
World Trade Organization (WTO) 7, 104,
    127, 129, 135, 137, 151, 152,
    157, 160, 162
    next talks round 133, 149, 161
World Wide Fund for Nature/World
    Wildlife Fund (WWF) 48, 57

zero option 54, 132
zonal programmes 118, 120, 122,
    123